THE ANTOFAGASTA (CHILI) & RAILWAY

THE STORY OF THE FCAB AND ITS LOCOMOTIVES

J. M. TURNER — R. F. ELLIS

This locomotive started out as a 2ft 6in gauge 2−8−2 built by Hudswell Clarke & Co. Ltd. in 1907 but was later extensively modified to metre gauge, including conversion to a 2−8−0 and major rebuilding of its tender. No. 140 was photographed in March 1939 at Antofagasta but soldiered on until the early 1960s.

B. Fawcett/J. Buckland collection

Designed by Donald Binns, Copyright 1992 Locomotives International/J. M. Turner/R. F. Ellis.
Published by Locomotives International, 50 Long Meadow, Skipton, North Yorkshire, England BD23 1BW.
Typesetting by Hanson Typesetting Services Ltd., 68 Haworth Road, Cross Roads, Keighley, West Yorkshire.
Printed by Clifford Ward & Co. Ltd., Bridlington, East Yorkshire.

A LOCOMOTIVES INTERNATIONAL NARROW GAUGE SPECIAL

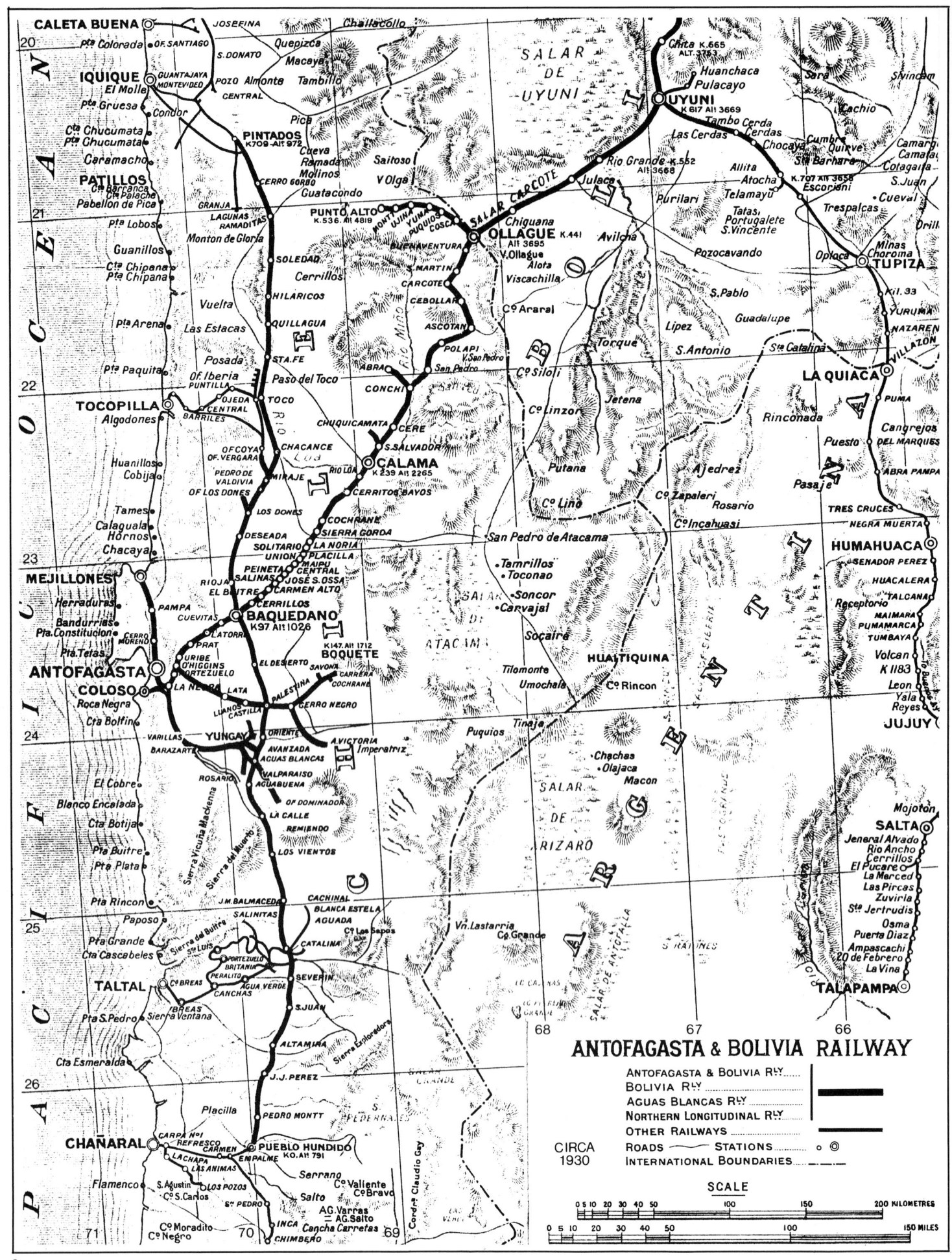

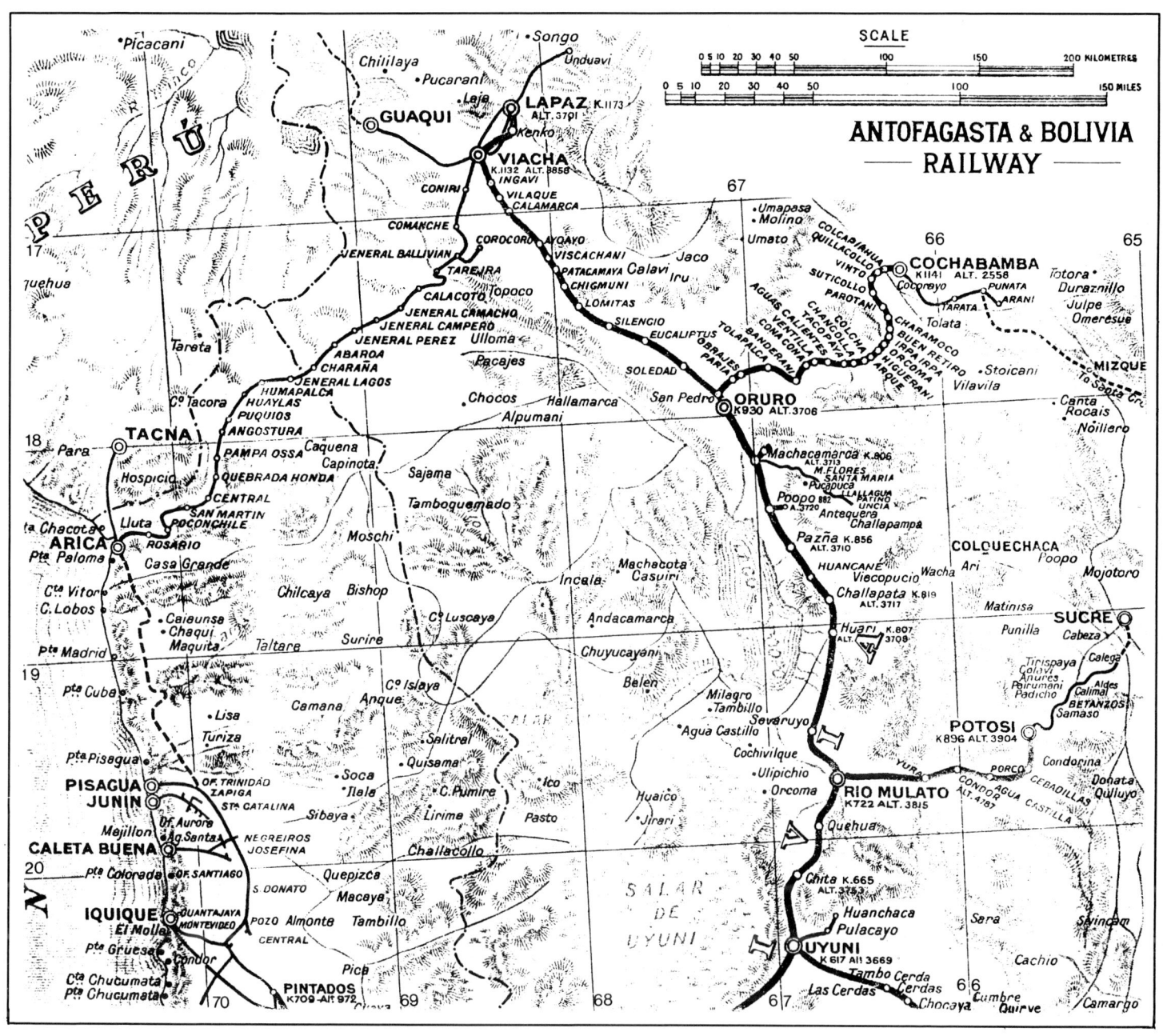

Outside front cover—top: The Old—this is one of the Baldwin 2-8-0s which became so popular on the FCAB narrow gauge. No. 79 "Portezeulo" built in 1906 was the leader of the final batch of this highly successful design tailored to Andean conditions. They had 15in×20in cylinders and 37½in driving wheels and of interest is the set forward cab, the tender cab covering and the unusual wrap-over sandbox. This engine was subsequently renumbered 113 and like other members of its class somewhat rebuilt in later years.   *H. L. Broadbelt collection*

Outside front cover—bottom: The New—No. 183 was one of a series of three 4-8-2Ts built by Hawthorn Leslie in 1929 and classified as "passenger engines". These very handsome locomotives were similar to the earlier standard North British 2-8-4Ts and shared the same boiler. Their business-like appearance included 21in×24in cylinders, 51in driving wheels, and oil fuel was carried in a bunker space above the cab. At 10ft wide, 12ft 10in high and weighing in at 86 tons they were quite large engines and depicted the later appearance of FCAB metre gauge power. The 4-8-2Ts were frequent performers on the Antofagasta—Ollague passenger trains.   *M. Swift collection*

CONTENTS	Page
1. THE FIRST 30 YEARS 1873–1903	4
2. CONSOLIDATION AND ACQUISITION	15
3. GAUGE CONVERSION	32
4. METRE GAUGE LOCOMOTIVES	35
5. THE FCAB CIRCA 1930	46
6. RECENT YEARS—CONTRACTION AND WITHDRAWAL	53
— FCAB NAME LISTING	56
— A PICTORIAL RECORD OF THE LAST 25 YEARS OF FCAB STEAM IN BOLIVIA	58
— BOLIVIA RAILWAY CARRIAGES AND WAGONS	70
— FCAB DISTANCES AND OTHER DETAILS	74
— ACKNOWLEDGEMENTS	Inside Back Cover

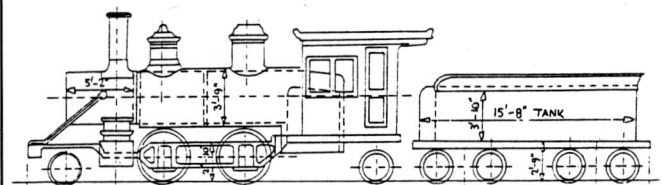

THE FIRST 30 YEARS 1873–1903

1988 marked the centenary of the formation of the Antofagasta (Chili) and Bolivia Railway Co. Ltd. or, as it is known in Chile, the Ferrocarril de Antofagasta a Bolivia. Apart from articles in the railway press in the 1910–1930 period and the odd chapter in books on South American railways in general, or Andean railways in particular, little has been written about this remarkable railway. At various times the line has laid claim to being both the world's highest and to having one of the most extensive 2ft 6in gauge networks. It is only when one studies the engines and rolling stock that one can appreciate the problems in moving close to 2 million tons of freight per annum by 2ft 6in gauge over a 12,000ft high mountain range!

Situated in Northern Chile the railway connected Antofagasta, an open roadstead port on the Pacific coast, with Ollague, 274 miles to the east at 12,123ft and on the border with Bolivia. On its way eastwards the line crosses the Atacama Desert, one of the world's driest regions.

A traveller to the railway described it in *Railway Engineer* for February 1891:

> "The line to Huanchaca was, for the most part, easy of construction. Leaving Antofogasta, the road goes up a canon to the tablelands. The track is laid directly on the bottom of the canon, with no embankment whatever. A two hours' rain would wash out a hundred miles of it; but there is no danger—it *never* rains there. The line crosses the desert of Atacama, passing through many valuable nitrate fields. There are said to be many rich mineral deposits in the broken ranges of mountains that dot the desert. It is to be hoped so, for a more God-forsaken country I never saw. For the first 300 kilometres, to the town of Calama, there is not a tree, not a blade of grass, nothing but rock sand and dirty white chunks of nitrate. Calama is an oasis—nothing more. It is the same thing beyond. There is no water the whole distance, and the stations have to be supplied with water drawn from Antofogasta, where water is obtained from the Pacific Ocean by distillation. The few mines that are worked on the desert obtain water from the nearest station. In some cases it is packed on mules a distance of forty miles. Although this road was easy of construction, it does not follow that it was built cheaply. Labour is scarce at from $2.50 to $5 per day, and supplies are very high. The country produces absolutely nothing. The road does a good business, bringing down nitrate and ore. The Huanchaca Company is putting in a $4,000,000 plant in Antofagasta, and will haul all its ore here to be treated. The gauge of this road is 30in., but it will undoubtedly be changed when connection is made with the Argentine system. As is the case with all roads to Chili, there are very few foreigners employed. The rolling stock is nondescript, but they have some good engines."

Historically the line dates back to 1872 when a concession was granted by the Government of Bolivia (at that time the Province of Antofagasta formed part of the Republic of Bolivia) to Melbourne Clarke & Company for the construction of a 2ft 6in gauge railway from Antofagasta to Melbourne Clarke's nitrate grounds at Pampa Alta. They then founded the Compañia de Salitres y Ferrocarril de Antofagasta (Antofagasta Nitrate & Railway Company) in Valparaiso, Chile, on 9 October 1872 and transferred the railway concession to that company. Rail was first laid in 1873 and the first section of 18.6 miles (29.9km) to Salar del Carmen opened on 1st December that year. Initially, mules were used as motive power to bring the caliche (nitrate bearing soil) down to the Pacific coast. The railway was extended to Salinas in 1879—the start of the Antofagasta nitrate pampa at 5,000ft above sea level. It was further extended to Central in 1882, and to Pampa Alto (km 150) in 1883.

Nitrate pampas in the Atacama Desert had been developed with Chilian enterprise, labour and capital. Punitive Bolivian taxes on nitrate were imposed in February 1878 and this heightened the tensions between Chile (whose citizens composed 90% of Antofagasta's population) and Bolivia. Bolivia was unwilling to arbitrate, as per the protocol of 21 July 1875 and Chile unwilling to let its citizens bow to Bolivian pressure. The resulting war broke out in April 1879, Peru sided with Bolivia, naval and land battles followed, with Chile becoming victorious. The War of the Pacific (as it was called) was settled with the Treaty of Ancon signed in October 1883 in which Bolivia ceded its littoral territories to Chile, whilst Peru lost its nitrate deposits in the Atacama Desert and southern province of Tarapacá.

Pampa Alta was still the end of the line on 17 January 1884 when the Chilean Government (which had taken possession of territory extending eastwards as far as Ollague) granted to the Compañia de Salitres y Ferrocarriles, a concession for an extension of the main line to Ollague (mp274) close to the newly established Bolivian border. Extension to Ollague proved inadequate since the roads of the interior that converged at that town could not provide enough traffic to feed the railway and it was decided that sufficient traffic could only be obtained by extending to Uyuni to tap the rich mining country around Pulacaya.

Another early development was the formation in 1873 of the Compañia Huanchaca de Bolivia, established to develop the valuable mining rights at Pulacayo in the Department of Potosi (Bolivia). Although successful from the start, the Company was handicapped by transportation problems and at first its mineral products were despatched to the port of Cogiba which required an 18 day journey over 480km of poor roads by animal drawn vehicles. Later, as a result of the War of the Pacific, it became necessary to find an outlet to the Atlantic Ocean through Rosario, Argentina. Because of this inconvenience the Compañia Huanchaca de Bolivia set its eyes on rail connection with the Pacific seaboard and the Compañia de Salitres y Ferrocarril de

Antofagasta was equally anxious to handle the resulting traffic, but was unable to span the intervening gap from lack of capital. In May 1885 the Compañia Huanchaca entered into agreement with the railway company to jointly build the extension to Ollague and push the line a further 120 miles to Uyuni in Bolivia close to the Huanchaca's silver mining operation. The Huanchaca Company would provide the necessary money to build the extension providing the railway company would put the whole of its line and equipment in the pool, as its capital contribution in kind, the railway to be operated for the joint benefit of the two undertakings. The project proved impracticable and the discussions were terminated by the Compañia Huanchaca purchasing all rights in the Compañia de Salitres y Ferrocarril de Antofagasta, for a sum of 3,000,000 paper pesos. This transaction was completed in 1887 leaving the Antofagasta company to concentrate on nitrate extraction. Huanchaca's only interest in the railway lay in its guaranteed access to a port and control over freight tariffs, without which its mining enterprise was not viable. This the Huanchaca company obtained by coming to London and offering to dispose of its recent acquisition. The possibilities for creating traffic were excellent and so the Antofagasta (Chili) & Bolivia Railway Company was floated on the London Stock Exchange in 1888, capitalised at £2,150,000 with the expressed purpose of acquiring from the Bolivian mining house the 2ft 6in gauge railway. On 27 November that year the transfer was finalised for the sum of £1,450,000 of which £1.15 million repesented the railway assets and £300,000 the value of the associated waterworks.

The Huanchaca Company retained the right to work the line for a 15 year period commencing on 1 January 1889 at 55% of the gross receipts and waterworks at cost price—thus freeing its capital for other purposes. The then President of Bolivia, Dr. Aniceto Arce, was on the Huanchaca board and had given the company a Government guarantee of 6% interest on any capital employed in the construction of the railway from Uyuni to Oruro. The line reached Uyuni (km617) in 1889 and Oruro (km929) in 1892 and the Huanchaca company smiled all the way to its bank.

In April 1889 the branch opened from Uyuni (on the main line) to Pulacayo and the silver mines of the Compañia Huanchaca de Bolivia, a distance of 20½ miles (33km). This branch, which included a 2 mile tunnel was never part of the FCAB, being retained and worked by the mining company. In

later years it was not converted to metre gauge, possibly due to restricted clearance in the tunnel.

Oruro was the end of the 2ft 6in gauge track but during its 15 year lease the Huanchaca Company added additional mileage to the system. Some 37 miles (60km) of sidings were laid on the Antofagasta pampa to serve the nitrate oficinas located between mp72 and 107 (km 117 and 170) at an altitude of 4900–5500ft. A 12¾ mile (20km) branch from Conchi to the copper mines at Conchi Viejo opened in February 1902.

The Huanchaca Company also constructed a 6¼ mile (10.08m) branch from the main line at San Salvador just northeast of Calama to the site of the great open cut copper mine at Chuquicamata—a branch which was to prove of the greatest importance to the FCAB. The original concession by the Chilean Government had been granted to a Mr. Norman A. Walker in July 1900, but the Huanchaca Company built the line and advanced the necessary finance at 8% interest, this to be repaid within two years. Agreement was reached in August 1900 for the Huanchaca Company to work the branch for 60% of the gross receipts and take an option to purchase it. These arrangements were subsequently modified, the option to purchase being set at six months within two years of the completion of the line with Walker receiving a quarter of the gross receipts for three years by way of compensation. In January 1902 the new branch opened and was purchased by the FCAB from the Compañia Huanchaca de Bolivia for £12,500 though the latter continued to work it until 1904.

For many years an absence of water had harassed the Company and in order to satisfy the thirst of its locomotives and other needs, water had been drawn from the Pacific Ocean from whence it was passed through huge condensers to distil it into fresh water. Eventually railway engineers at work in the mountains discovered spring water at 14,500ft. Pipes were installed to bring the water down the mountain for 40 miles to San Pedro station (10,607ft) where this was piped into storage reservoirs blasted out of the mountain sides. An eleven inch pipeline was then laid parallel to the railway for most of its 197 miles to Antofagasta and could handle 6000–7500 tons daily. This facility served the railway, the nitrate fields and the city of Antofagasta. The project cost some £1,500,000. *Stock Exchange Official Intelligence* for 1905 says "The waterworks consist of 342km of aqueducts bringing the water from the Polapi Springs and the River San Pedro to Antofagasta".

The entire system passed into full British control at midnight on 31 December 1903. To date we have not sighted a Return for Engines and Rolling Stock transferred to the FCAB at this time, although it seems certain that the vast majority of the locomotive stock of the two earlier companies was still in existence. This included a family of R. Stephenson built 4–6–0s of 1876 (with a repeat order in 1887–8), some small Avonside and Sharp Stewart 0–6–2Ts and 2–6–2Ts, four 2–4–2s, one from Rogers and three from Baldwin, a Rogers 0–4–4T Inspection Coupe, and a large family of Baldwins dating from 1889 which covered a number of wheel arrangements: 2–6–0, 2–8–0, 4–6–0 and 0–6–2ST, some of which were tender/tank engines. There was also a single Cail 4–8–0 of 1898. We do know however, from an article in the *Railway Magazine* for August 1909 that some of the 1876 R. Stephenson 4–6–0s were still in existence.

Three axle mineral wagon photographed at Mejillones Works April 1985, date built unknown, but in all probability a 2ft 6in gauge pre-1904 survivor, used on the Aguas Blancas, until its closure in 1961. I. Thomson

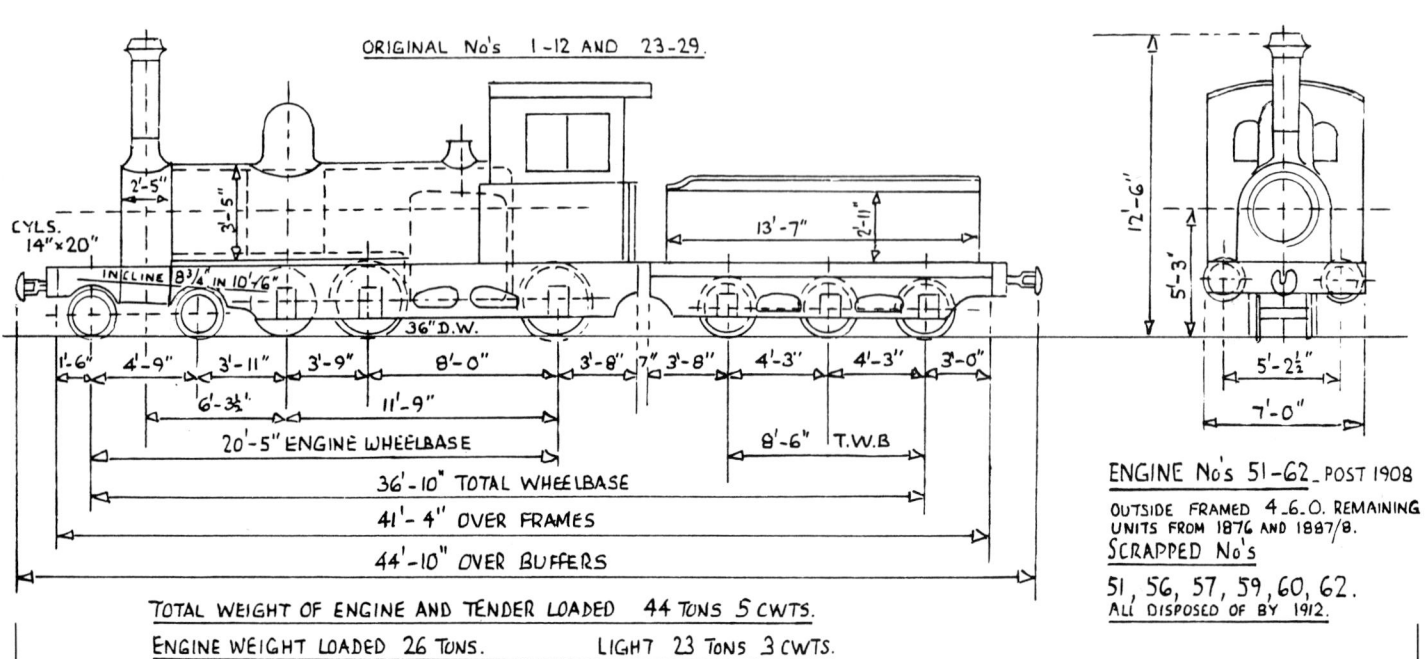

ANTOFAGASTA (CHILI) AND BOLIVIA RAILWAY CO. LTD.

LOCOMOTIVE LIST

1. **Old Numbering Series for 2ft 6in gauge locos from 1876 to 1908.**

   This period covers the two constituent companies:
   Compañia de Salitres y Ferrocarriles de Antofagasta (1876 to 1887) and the
   Compañia Huanchaca de Bolivia (1887 to 1903) until the general renumbering programme in 1908.
   The majority of early locomotives were named, but at times these were changed.
   There is also some doubt about the authenticity of names and when they were carried. To keep the lists within a reasonable length the names have therefore been omitted.

1.1 Compañia de Salitres y Ferrocarriles de Antofagasta

Original Numbers	1908 Numbers	Type	Builder	Works Numbers	Year Built	Remarks
1–12	–	4–6–0	RS	2291–2302	1876	
13–14	–	0–6–2T	AE	1182–1183	1877	
15	1	0–6–2T	AE	1195	1877	
16	–	0–6–2T	AE	1196	1877	
17	2	2–6–2T	SS	3032	1882	Rebuilt as 0–6–2T prior to 1912.
18–19	–	4–2–4–2T	RS	2449–2450	1884	Webb compounds rebuilt to 4–6–0s at an unknown date.

1.2 Compañia de Huanchaca de Bolivia

Railway officially known as Empresa del Ferrocarril de Antofagasta de Bolivia after purchase by the above company in 1887.

Original Numbers	1908 Numbers	Type	Builder	Works Numbers	Year Built	Remarks
20	33	2–4–2	BLW	8215	1886	Renum. 32 1907/8, then 33 after withdrawal of original No. 40. This loco was converted to metre gauge at a date unknown, renum 351, and noted as such on the FC Bolivia by B. Fawcett in 1939.
21	42	2–4–2	Rogers	3713	1887	
22	See note	0–4–4T	Rogers	3709	1887	Inspection Coupe disposed of prior to 1912. Shown in Diagram Book c.1908 numbered 31 and in a photo taken 10/1903 numbered 33. Rebuilt to a conventional tank engine prior to 1908.
23–25	–	4–6–0	RS	2622–2624	1887	
26–29	–	4–6–0	RS	2633–2636	1888	

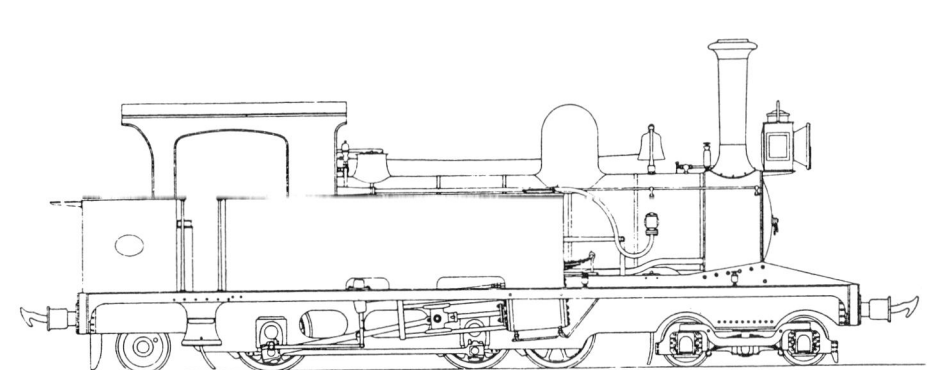

4–2–4–2T Robert Stephenson & Co. Ltd. Works No's 2449–50 of 1884. Webb system of compounding. Two high pressure cylinders 10in diameter × 20in stroke, outside the frames drove two coupled axles, whilst the 20in diameter × 18in stroke low pressure cylinder placed between the frames was connected to an uncoupled axle located between the leading bogie and the rear group of driving wheels.

Leading and trailing bogie wheels	2ft 0in diameter		H.S. Tubes	634 sq. ft.
All driving wheels	3ft 0in ,,		H.S. Firebox	61 ,, ,,
Bogie wheelbase	4ft 9in		H.S. Total	695 ,, ,,
Coupled wheelbase	8ft 0in		Grate area	11½ ,, ,,
Total wheelbase	26ft 1½in		Water	550 gall.
Boiler diameter	3ft 5in		Wt in w.o.	about 35 tons
Boiler length	10ft 10½in			
Firebox	4ft 6in long × 3ft 6in wide		Later rebuilt to 4–6–0s.	

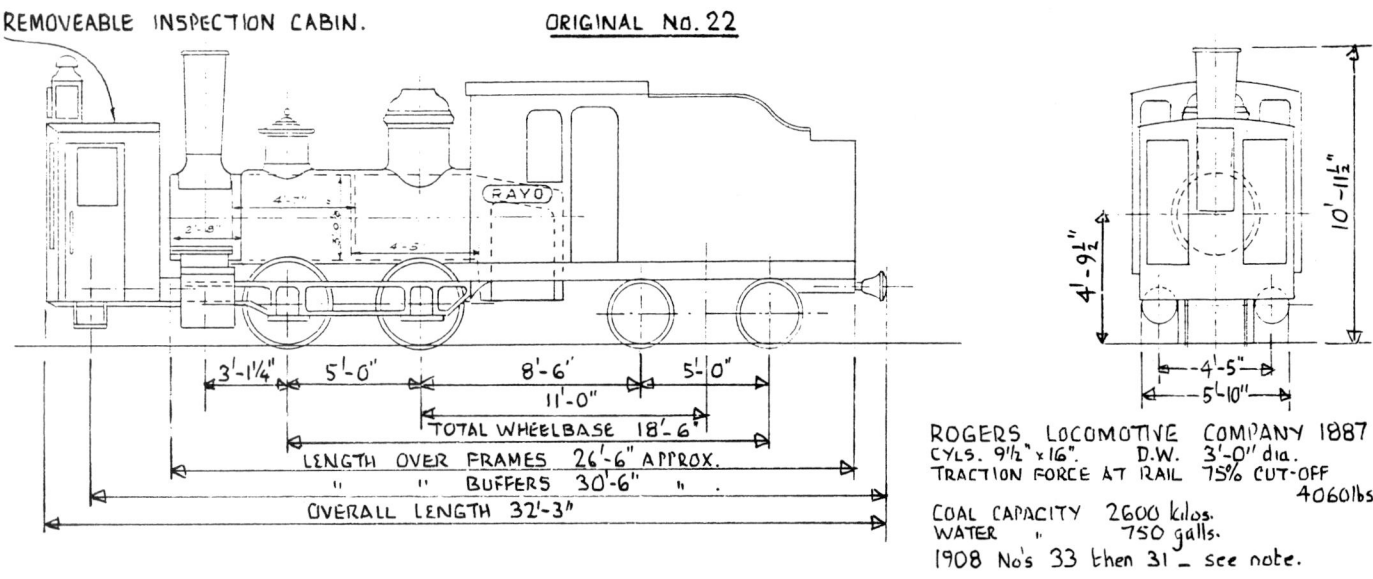

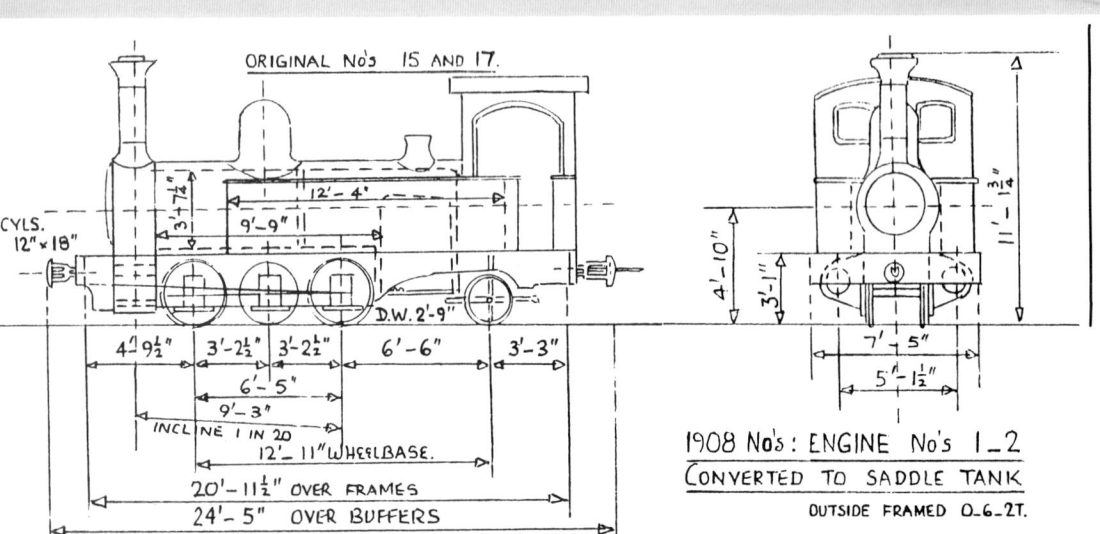

Rogers Locomotive Company 0-4-4T Works No. 3709/1887 was photographed in October 1903 with inspection cabin mounted ahead of the smokebox. Note the elaborate lining applied to this locomotive then numbered 33, but subsequently renumbered 31.
R. N. Redman collection

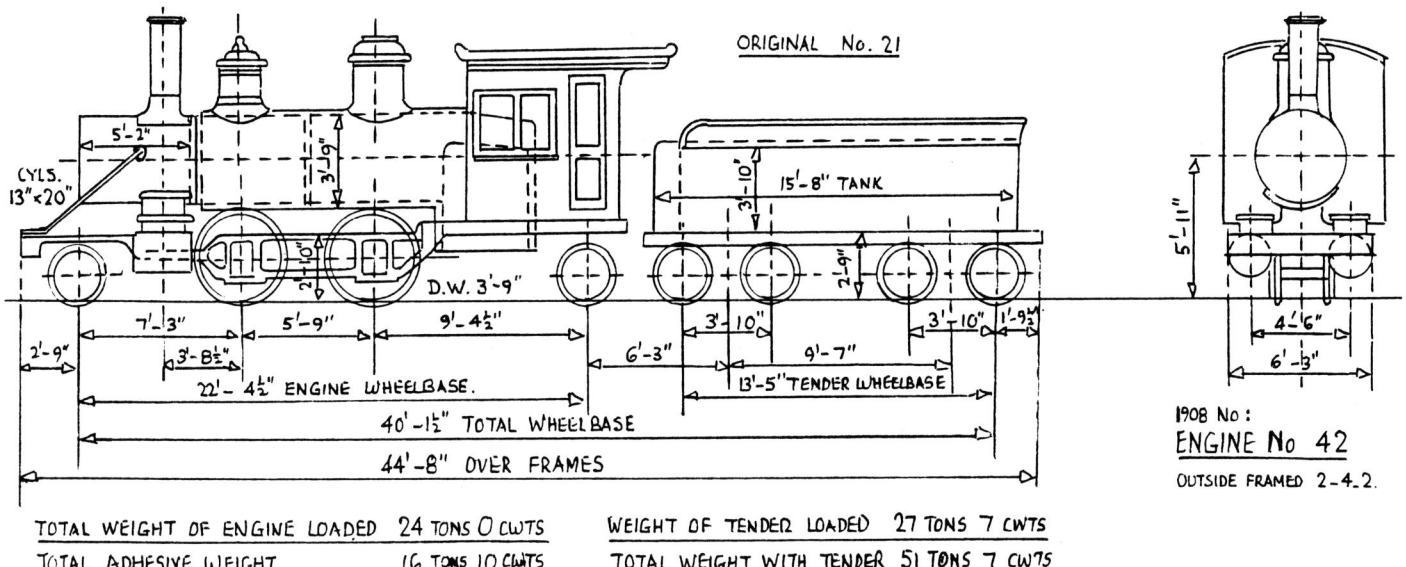

Railway sold to the Antofagasta (Chili) and Bolivia Railway Co. Ltd. on 28/11/1888, after which the Huanchaca Company leased the line for a period of 15 years, terminating on 31/12/1903. During this time the Huanchaca Co. used several different business names when purchasing locomotives.

Original Numbers	1908 Numbers	Type	Builder	Works Numbers	Year Built	Remarks
30	71	2-6-0TT	BLW	9846	1889	Reported Scr. in 1916 report.
31	72	2-6-0TT	BLW	9852	1889	
32	82	2-6-0TT	BLW	9855	1889	
33	83	2-6-0TT	BLW	9864	1889	
34	84	2-6-0TT	BLW	9859	1889	Converted to metre gauge in 1917. Out of service by 1925.
35	95	2-8-0TT	BLW	9773	1889	Scr. in 1916 report. Side tanks removed prior to 1907/8.
36	3	0-6-2ST	BLW	9770	1889	
37	81	2-6-0TT	BLW	10469	1889	
38–39	77/76	2-6-0TT	BLW	10470/10464	1889	
40	–	2-4-2	BLW	10942	1890	Renum. 33 1907/8. Scr. in period 1908–10.
41–42	34–35	2-4-2	BLW	10943–10944	1890	
43	73	2-6-0TT	BLW	10984	1890	
44	74	2-6-0TT	BLW	10988	1890	
45	75	2-6-0TT	BLW	10997	1890	Scr. in 1917 report.
46	–	0-6-2ST	BLW	10998	1890	"Sold" to Huanchaca Co. c.1892 for use on the FC Uyuni–Pulacayo and possibly initially used on construction of the line.
46 (2nd)	see No. 51 below and notes					
47	4	0-6-2ST	BLW	10995	1890	
48	78	2-6-0TT	BLW	11426	1890	
49	79	2-6-0TT	BLW	11436	1890	
50	80	4-6-0TT	BLW	11437	1890	Listed in BLW data as 2-6-0TT but p.81 of *The Locomotives that Baldwin Built* by Fred Westing (1966) illustrates an unlettered 4-6-0ST&T for the FCAB and company diagram book records it as a 4-6-0TT.
51	7	0-6-2ST	BLW	12752	1892	Renumbered 2nd. 46—see notes below.
52	8	0-6-2ST	BLW	12753	1892	Renumbered 2nd. 51—see notes below.
53	9	0-6-2ST	BLW	12754	1892	Renumbered 2nd. 52—see notes below.
53 (2nd)	96	2-8-0	BLW	12635	1892	
54	97	2-8-0	BLW	12633	1892	
55	98	2-8-0	BLW	12667	1892	
56–57	10–11	0-6-2ST	BLW	14220–14221	1895	
58–60	92–94	2-8-0	BLW	14461–14463	1895	58 unserviceable in 1917 report.
61–62	43–44	4-4-0	BLW	14464–14465	1895	Both scr. in 1916 report.
63	91	4-8-0	Cail	2466	1898	Shown in some lists as 4-6-2: company diagram shows 4-8-0 is correct.
64–65	99–100	2-8-0	BLW	17461–17462	1900	65 scr. in 1917 report.
66	–	2-8-0	Rogers	5544	1900	Retained by the Huanchaca Co. in 1903.

Original Numbers	1908 Numbers	Type	Builder	Works Numbers	Year Built	Remarks
67–70	101–104	2-8-0	BLW	18388–18391	1900	101 scr. in 1917 report.
71–74	105–108	2-8-0	BLW	19437–19440	1901	108 last loco ordered by the Huanchaca Co.

There is some confusion with the running numbers in this section as BLW numerical registers record 52–62 with higher running numbers than shown. About 1892, when No. 46 went to the Huanchaca Co., a new Baldwin No. 51 was renumbered 46 to fill this gap and the other engines delivered at this time were renumbered accordingly. It appears that Baldwin were not notified of this change, and as a result allocated higher running numbers to the later locos. It is unclear from available information whether these were given their correct numbers on arrival in Chile though this seems possible. To confuse the issue it had long been assumed that the running numbers were sequential with the Baldwin construction numbers, i.e. 46 = 10995 and 47 = 10998. However, this does not appear to have been the case. The information on the renumbering of 51 to 46 is confirmed by the company's diagram book from the period 1908–1930. They had a curious habit in the pre-1912 period of using running numbers in lieu of builders' numbers, so the solution had to be found using names and dates. An early Livesey & Henderson list (the company's consulting engineers) confirms 47 as 10995, and a 1955 visit by a party of enthusiasts to Huanchaca confirmed that a loco numbered 46 was still in service. Two separate Baldwin lists give conflicting information, but both confirm that the unit transferred carried the same name but quote different builders' numbers.

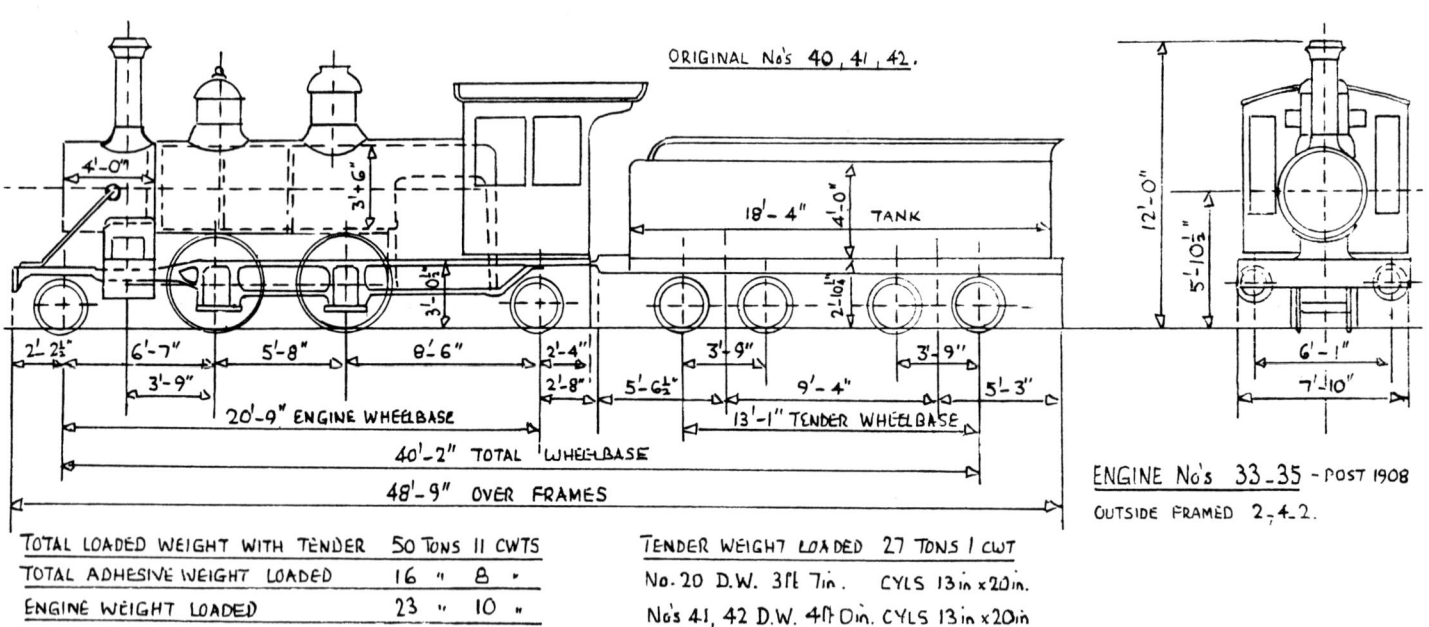

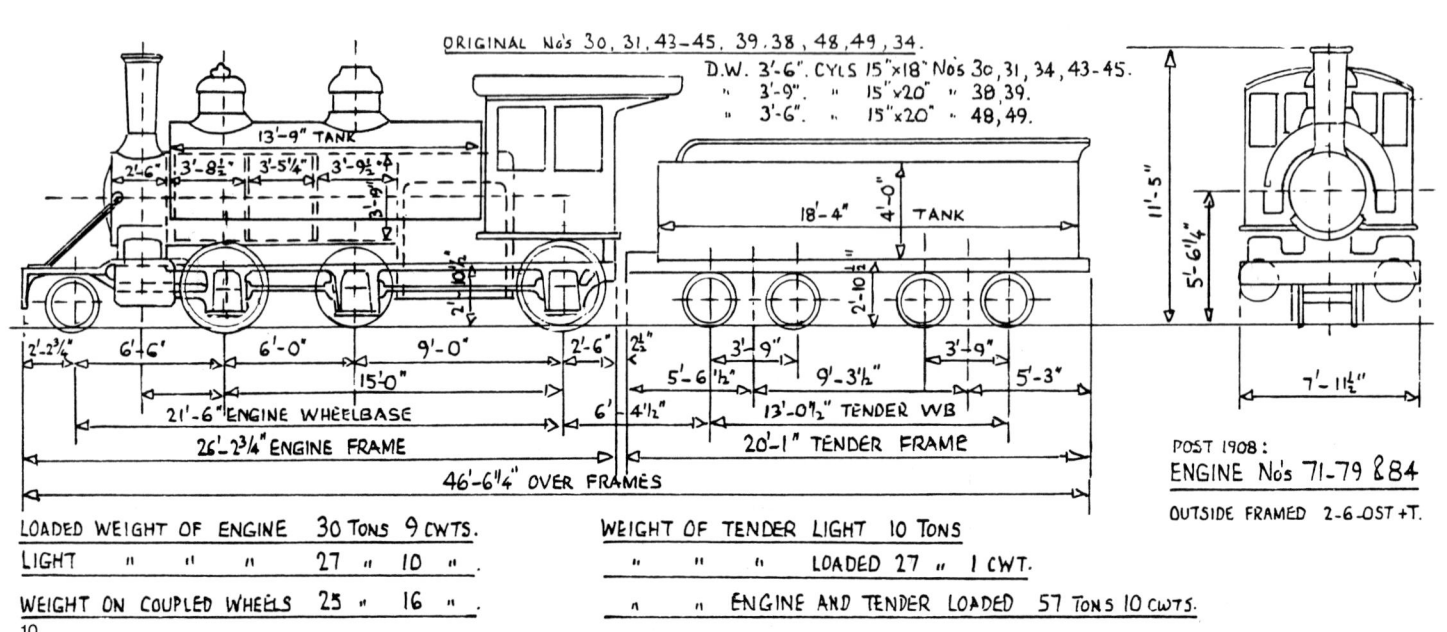

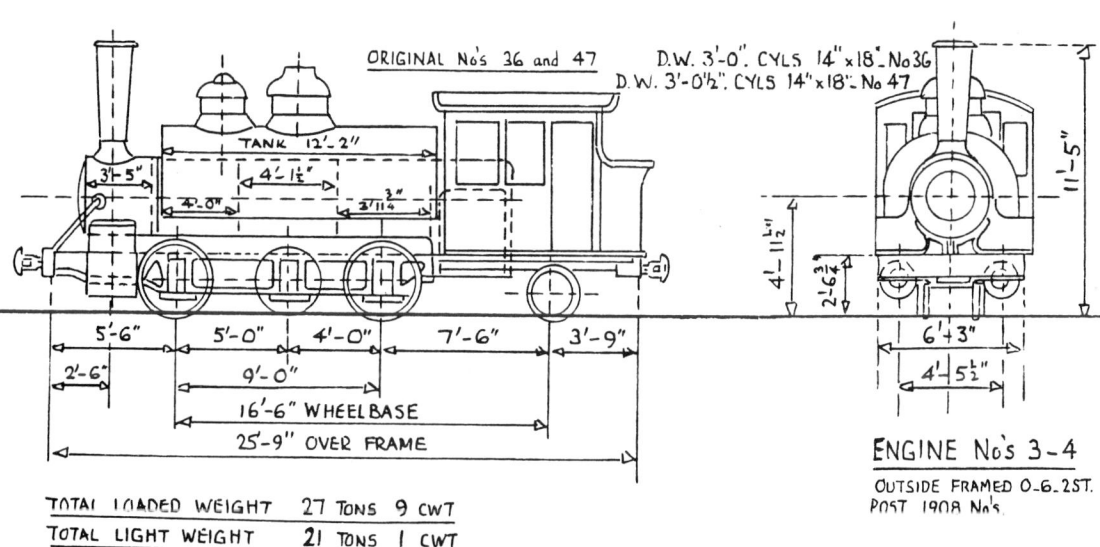

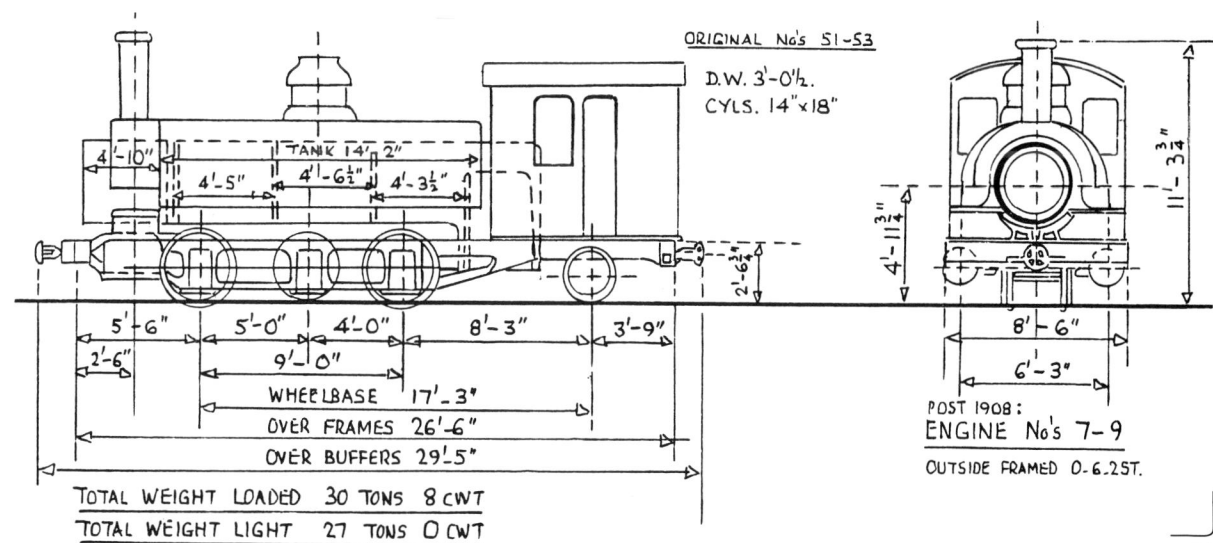

In addition to its six and eight coupled main line locomotives, the FCAB also had a family of 0–6–2STs on the narrow gauge which were used for shunting. No. 36 "Relámpago" built by Baldwin in 1889 as a single engine order was one of them. It was a very conventional American design with 14in × 18in cylinders, 36in driving wheels and an enormous oil headlamp that looks like it really belongs to a standard gauge loco!

H. L. Broadbelt collection

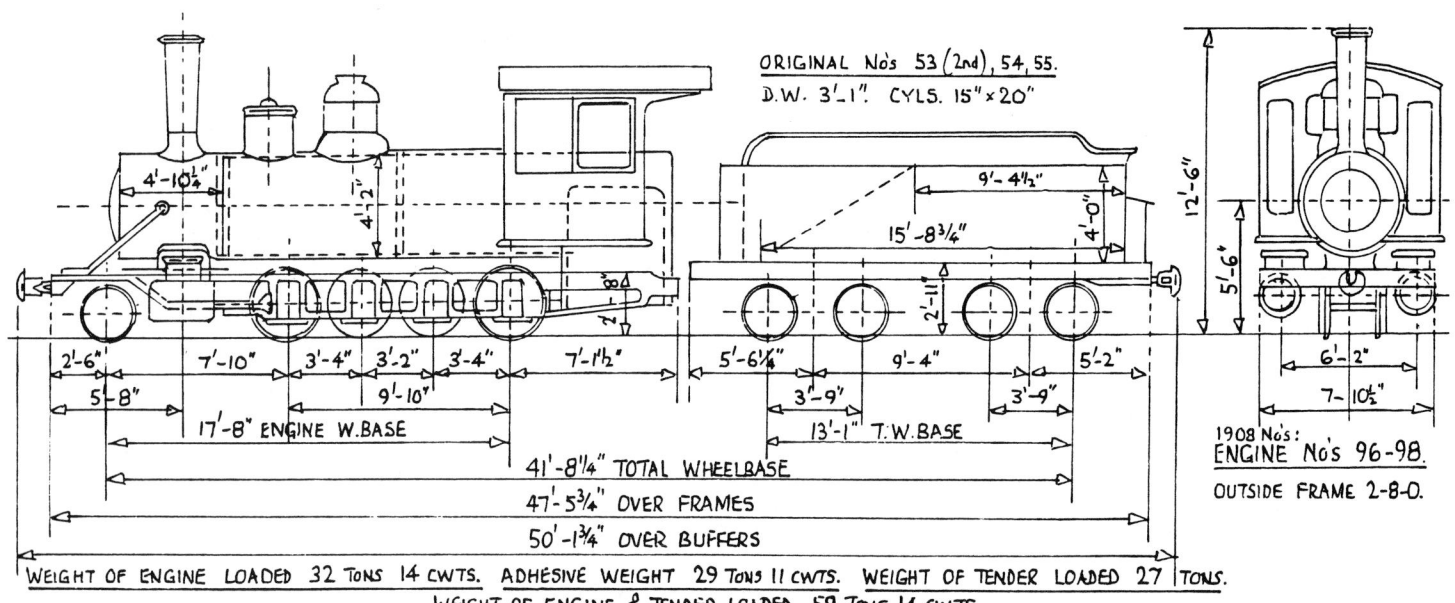

"Vulcano" was one of a batch of three Baldwin-built 2–8–0s and carried Works No. 14462 of 1895. It was typical of many little "Consolidations" to be seen on the 2ft 6in gauge tracks of the FCAB and its predecessors. Note the cast name and number plates on the cab-side and the large painted number "59" on the tender. "Vulcano" was photographed in the period 1895–1907.   C. J. Walker collection

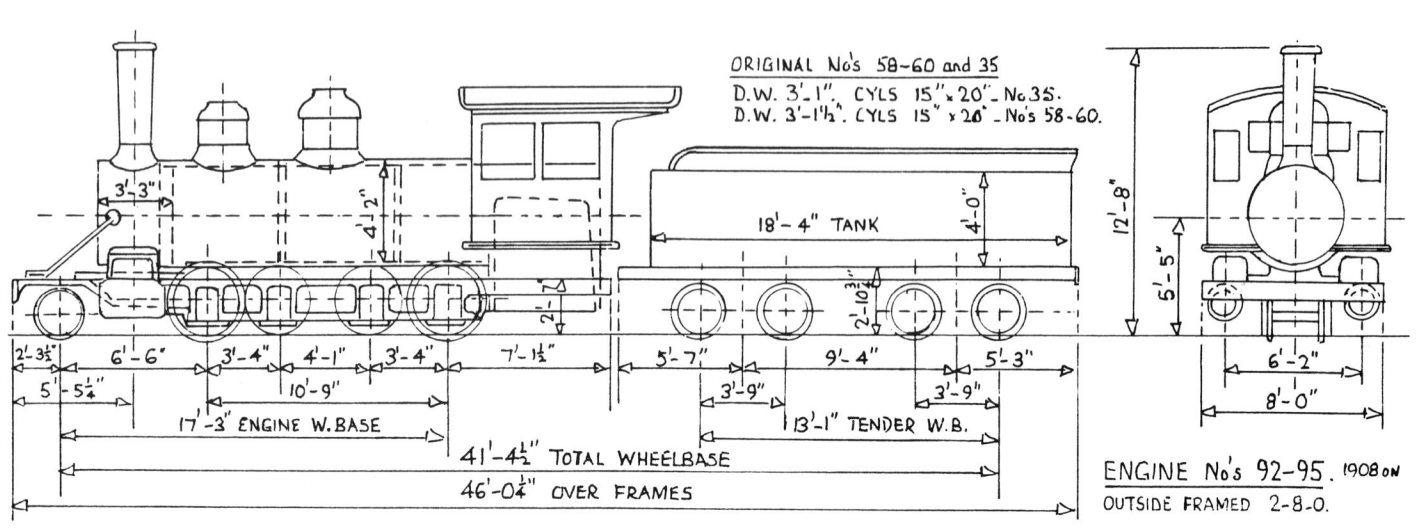

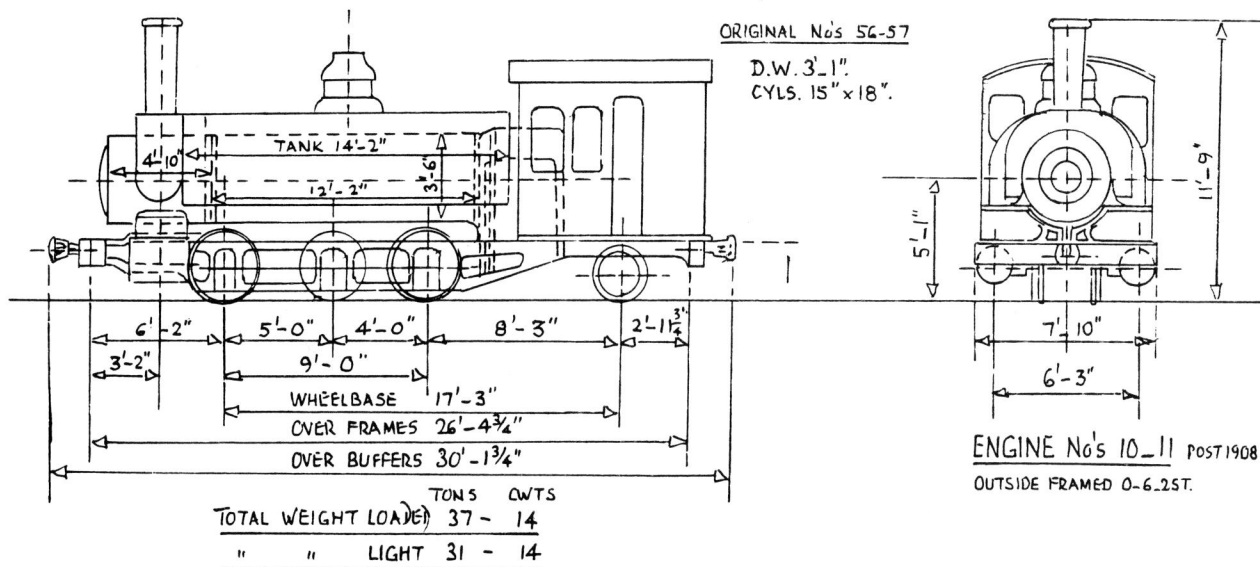

Passenger locomotives were rare on the FCAB on both gauges, the preference being for mixed traffic designs, generally eight coupled. 4–4–0 No. 62 with its long wheelbase was one of two built by Baldwin in 1895. It had the standard 15in × 20in cylinders, but 43in driving wheels. The 4–4–0s appear to have been of limited use as they had both gone by 1916.
H. L. Broadbelt collection

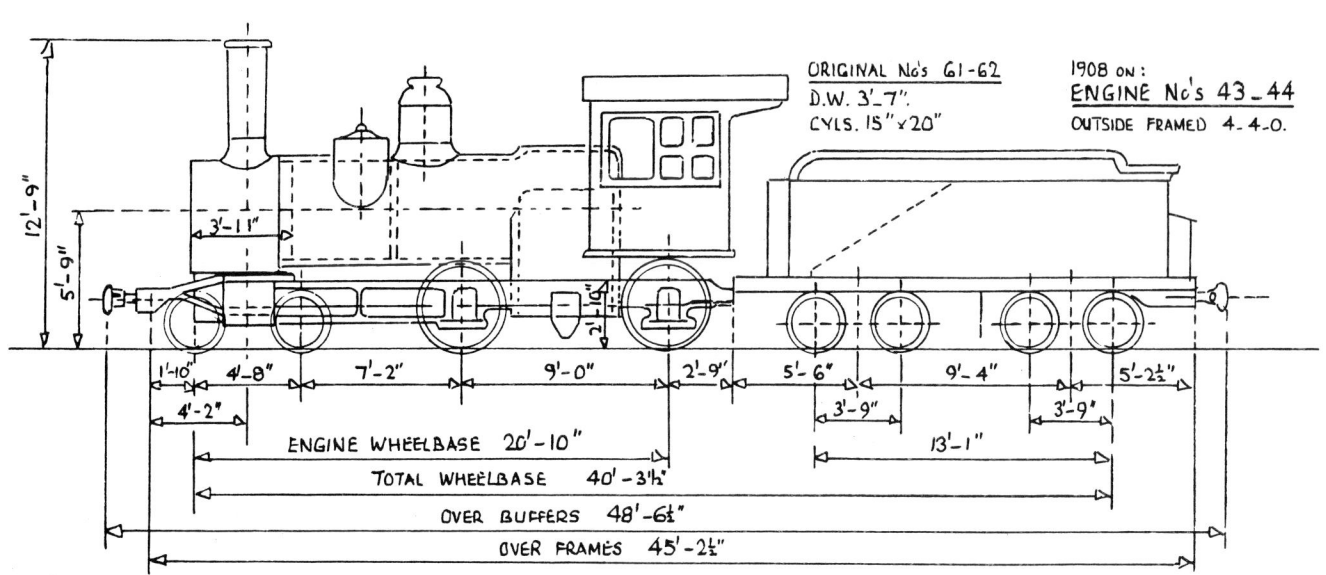

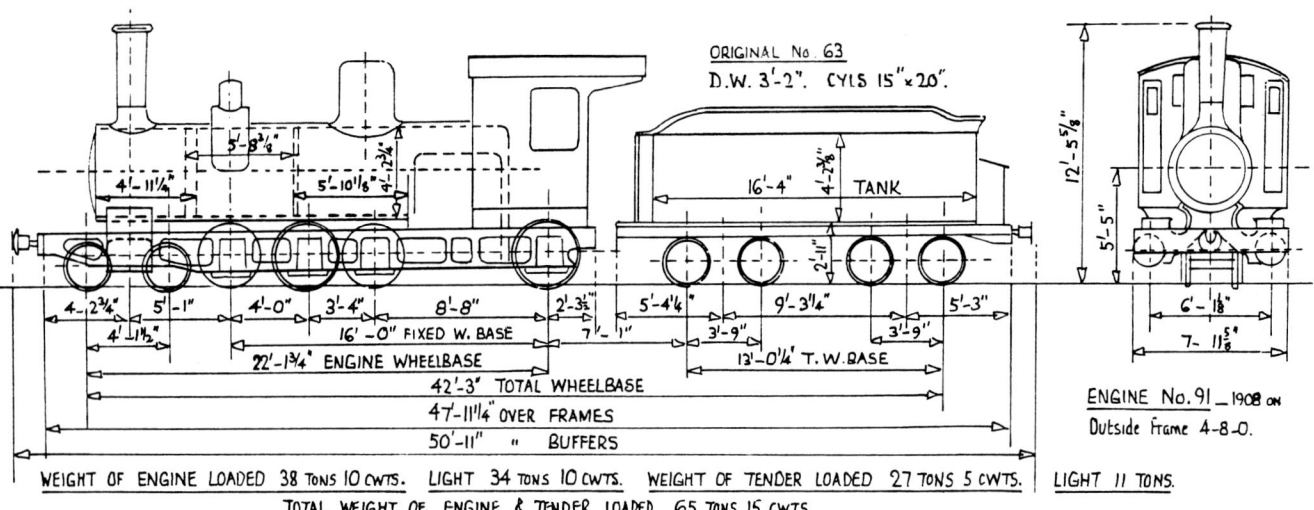

2–8–0 No. 66 "Porvenir" was built by the Rogers Locomotive Works, Paterson, New Jersey in 1900 and arrived soon after a large family of Baldwin 2–8–0s suggesting it was purchased for comparison, It must have been successful for other locomotives followed. The engine has a typical American appearance with outside bar frames and the boiler projecting right back into the cab.
Alco Historic Photos

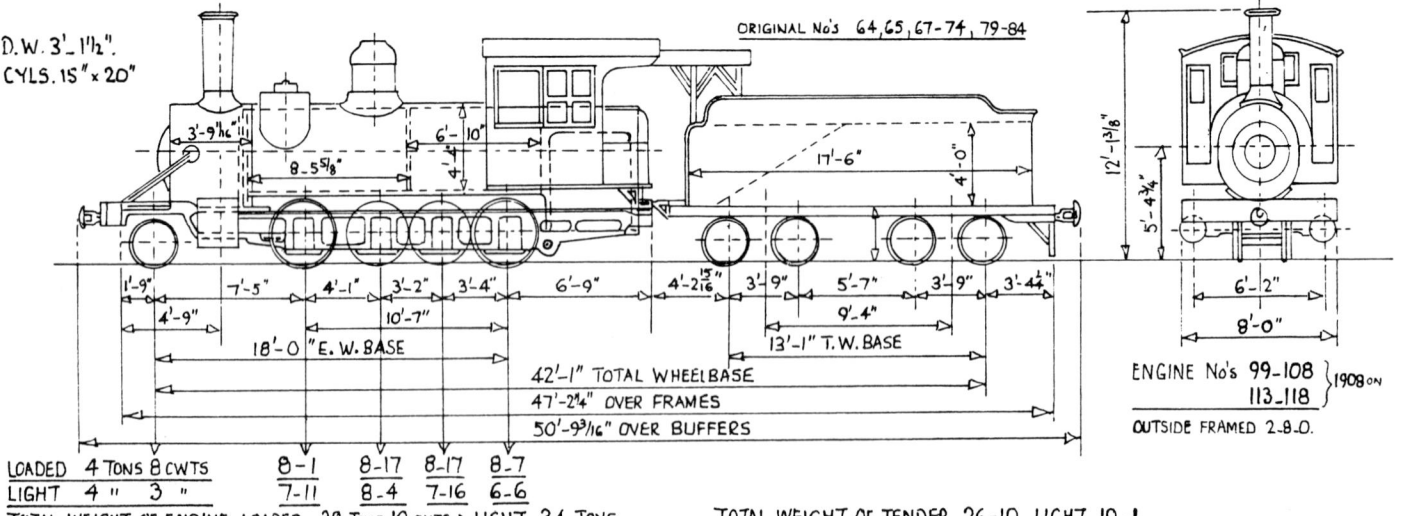

1.3 Ferrocarril Uyuni–Pulacayo

After termination of its lease in 1903 the Huanchaca Mining Co. continued to operate its line from Uyuni to Pulacayo and Huanchaca. The line remained 2ft 6in gauge and continued to operate well into the 1970s under the above name. With two exceptions previously mentioned and shown below, the number of locos retained by the Huanchaca Co. to work this line has not so far been found. The rest of the locos shown below were direct purchases by the company and were not numbered in the main line register.

Original Numbers	Type	Builder	Works Numbers	Year Built	Remarks
1 (?)	2-6-0	BLW	12363	1892	Ordered for the Pacamayo & Huanchaca Railway.
2-3	0-4-0ST	BLW	12404-12405	1892	
?	0-6-0ST	BLW	13997	1894	
4	0-4-0ST	BLW	14301	1895	Baldwin records list: "Huanchaca Co.–Corp. Minera de Bolivia No. 4". Observed in 1955 with plate: No. 1 "Playa Blanca".
5	2-6-0	HL	2947	1912	Observed in 1955—rebuilt from FCAB No. 165. This would have required a drastic boiler shortening, more likely a new, shorter boiler was supplied.
12	4-6-0	HE	?	?	Observed in 1955—this is one of the well-known ex-WDLR Hunslet 4-6-0s rebuilt into a tender engine. The identity of No. 12 is not confirmed. A total of five former WDLR 4-6-0Ts went to Chile, all to the order of Beverly Pearce & Partners. None had seen active service in the war, having been built by Hunslet in 1919 and regauged from 60cm to 2ft 6in prior to despatch to Chile. HE numbers 1367 (WDLR 3251) and 1374 (3258) were despatched in March 1920 to Antofagasta; 1357 (3241) in April 1920 to Antofagasta, and 1359 (3243) and 1368 (3252) to Iquique, also in April 1920. It is possible that all of the Antofagasta locos may have ended up here.
46	0-6-2ST	BLW	10998	1890	Main line loco retained by the Huanchaca Co. Named "Union". Observed in 1955 with same number confirming no renumbering took place.
66	2-8-0	Rogers	5544	1900	Main line loco retained by the Huanchaca Co.

The authors are aware that this list of locos which operated here may be incomplete and would welcome further information.

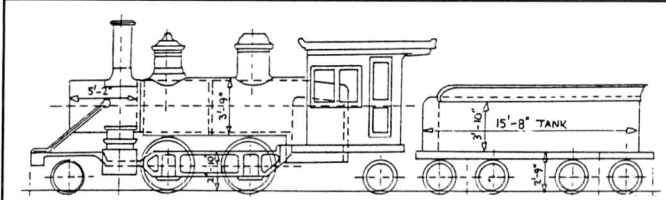

CONSOLIDATION AND ACQUISITION   2

From Ollague the 60 mile (95km) Collahuasi branch opened in April 1908 to serve what were among the richest known copper mines in the world. This line reached 15,795ft above sea level at Punta Alto which at the time was the highest rail summit in the world.

The resources of the port of Antofagasta were unable to cope with the growth in traffic brought about by rapid development within the copper and nitrate industries and it was decided to open up the port of Mejillones. A concession was obtained to build a 47 mile (76km) branch from Prat (mp34, km59) on the main line, via Pampa to Mejillones, this being constructed between 1904 and 1906, the new port facility becoming operational in that year. Most of this section up the coastal escarpment was through solid rock and involved sharp curves, deep cuttings, extensive fill, and 10 miles of 3% grades. Mejillones is one of the few natural harbours on the nitrate coast and wharfage was constructed to allow direct cargo handling. At the other nitrate ports there were open "roadsteads" which required tugs and lighter to transfer cargo to and from the ships anchored off shore. Mejillones was also the site of the FCAB's new workshops, building of which commenced in 1908. After several years of negotiations commencing in 1906, a concession was granted to lay a line running parallel to the coast north of Antofagasta for 22 miles (35km) to Pampa, thus making obsolete the expensive Pampa-Prat climb. The new line connecting Antofagsta with Pampa opened in 1914 and by 1919 the Pampa-Prat portion of the Mejillones route had been lifted.

FCAB locomotives and rolling stock which followed were built to the design and inspection of the Company's consulting engineers, Livesey, Son & Henderson, with one or two notable exceptions. The Antofagasta based operating management were initially suspicious of untried U.K. designs and when an order was placed with the Hunslet Engine Co. for delivery in 1906 for 4 2-8-2s to a new specification, a second order was placed with the Baldwin Locomotive Works for 6 2-8-0s to the previous successful Huanchaca design and specification. This type had been perfected over five different batches built by Baldwin from 1892, and was a development of a single Baldwin 2-8-0TT (Works No. 9773) supplied in 1889. The type was obviously popular as Rogers, and its successor—the American Locomotive Co.—supplied locomotives to the FC Caleta Coloso a Aguas Blancas built to the same drawings (Rogers Works No's 5701-5702/1902, Alco 38445/1905 41115-41116/1906).

A second batch of 20 2-8-2s basically duplicates of the initial order (first: piston valves, second: slide valves and redesigned boiler) was ordered in June 1906 to FCAB specification (Spec. F787 drawing 6646/7) with the order equally split between two British manufacturers, Hunslet and Hawthorn Leslie. An additional two locomotives were ordered from Hudswell Clarke & Co. but were modified to test the suitability of 39in drivers. This overall design was not successful, and the question of the suitability of the "Mikado" type was raised.

The boom in freight required that further orders be placed,

the hunt for higher horsepower ruled out the lightweight Baldwin with its 15in × 20in cylinders. Twenty 2−8−0 "Consolidations" were ordered from the American Locomotive Co. to a successful Alco narrow gauge design with 16in × 20in cylinders and 37½in drivers, and an order for ten to a new U.K. design, correcting the faults of the earlier 2−8−2s, but with 16½in × 20in cylinders, was placed with Hunslet, both orders for 1908 delivery.

To their credit Hunslet sent a factory representative, Edward Watkin, to Antofagasta in 1907/8 and his reports together with those of the Locomotive Superintendent make fascinating reading. The arrival of the 30 1908 "Consolidation" locomotives enabled the laying up of the earlier "Mikados" and they were little used until 1918. Of the 1906 batch, two were sold to the Boquete Nitrate Company prior to 1912, and the other two were scrapped in 1917/18. The defective boiler design of this initial batch mitigated against them being rebuilt.

American locomotives were flexible, and set a standard for the rough, unballasted Chilian sector track. The cylinders and smokebox saddle were cast as one unit, this, together with bar frames, and fully compensated soft springing, absorbed the punishment without placing additional strain on other components of the locomotive. Although the American locomotives were roughly finished, they were designed for ease of maintenance. Bolts were driven from the inside so as to expose the nut for inspection and ease of tightening. Pipework was accessible, unions were used throughout on all the air pipes. Bearings and crankpins were, to British eyes, over-large, and the tenders, riding on "diamond" bogies, the result of years of experience with atrocious track.

The FCAB opinions can best be illustrated by quoting extracts from letters received by Hunslet. 4 May 1907 from John Brown, Locomotive Superintendent. "Impossible to keep steam and exhaust joints from leaking under smokebox on account of the movement of framing. Water leaks from each side of framing and causes slipping. This would not be the case if the cylinder castings were the same as American design. Tender too well fitted and continually jams." 13 May 1907 from General Manager, "As indicated in my telegram, the Baldwin Engines constructed in America are better adapted for the type of road common in America than the Hunslet, which are built for the rigid roads common in England, and the modifications that Hunslet suggested are calculated to assimilate them to the Baldwin type. The workmanship of the English engines is far superior to that of the American, and if English makers would construct their engines on American lines they would be preferable."

Again from John Brown dated 7 May 1907, "the Baldwin engines received last year are generally satisfactory and infinitely more suitable for our line than any English engine supplied up to the present beyond the question of tubes (on account of the quality of the water). We have had no trouble with these engines at all and even the tubes leak a great deal less than with the English engines on account of the greatest elasticity of spring arrangement and framing."

Major Hilary Hood, the Chilian sector Locomotive Superintendent from 1918 to 1933 notes in a memo dated 14 August 1923 of "complaints of the permanent way department as to the severity of the initial 2−8−2 engines on the road and the name they earned — terra-motors (earthquakes)." These difficulties whilst mitigated as a result of Hunslets investigations in 1907-8 undoubtedy had a bearing on their popularity. Major Hood, when taking up his appointment found them little used, so rebuilt the entire class to 2−8−0s, adding a heavy cast iron beam to the leading end, and in this configuration they were noted later as giving excellent service, though by this stage the main line was fully relaid with heavier rail.

Far more successful was the introduction during this period of a series of British built tank engines. These included a group of 0−6−4Ts from Hunslet in 1905−7 and some 2−8−4Ts from Kitson in 1912. The 0−6−4Ts were used for shunting on the FCAB and locos to the same design were also built by Hunslet for the Chilian Longitudinal Railway where they were fitted with tenders to increase their availability. The 2−8−4Ts were long and narrow and had a rather racy look and were also used for heavy shunting. A number of the 0−6−4Ts were later extensively rebuilt as 0−6−0Ts and 0−6−2Ts and all these plus the 2−8−4Ts and remaining 0−6−4Ts (with the exception of four 0−6−4Ts which went to the Aguas Blancas line) were converted to metre gauge. In 1928 three of these oil fired locomotives were noted as being fitted out for one man operation. This was achieved by having the oil flow into the firebox controlled by the movement of the regulator — quite an ingenious way of overcoming the problem!

In an attempt to obtain increased power, Livesey, Son & Henderson experimented with the Meyer form of articulation in their designs. It is interesting to speculate why they did not choose the Garratt design which by this time was fairly well established (and in fact the FCAB did buy some Garratts later). One can only assume that Meyer design was chosen because of its already established reputation for railways in the Andes. One of the Meyer designs was for a 2ft 6in gauge 2−6−0+0−6−4T which consisted of three bogies of which two were driven. The cylinders were on the outer end of the first and second steam bogies and the tender was articulated between the second (middle steam bogie) and a trailing four wheel truck. Only two engines of this type were built — the one for the FCAB in 1908 and the other for the Leopoldina Railway in Brazil (to whom Livesey, Son & Henderson were also consulting engineers). The FCAB's second Meyer was a 2ft 6in gauge 2−6−0+0−6−2T delivered by Kitson in 1912 to a standard Kitson design. The third Meyer type purchased was, like the first, of Livesey, Son & Henderson design with Beyer Peacock as the successful tenderer as by this time Kitson were becoming cautious of outside Meyer designs. One wonders if Beyer Peacock were chosen to placate them for not using the Garratt type! Six metre gauge 0−6−2+0−6−2 tender locomotives were delivered in 1913 which were designed to run cab forward as a sort of double "mogul" type. These were very successful, particularly when converted to oil firing shortly after their arrival in Bolivia. They were quite long lived, passing into Bolivian National hands in 1964, and the remains of five were still in existence in May 1978 in the locomotive dump near Uyuni.

Loco tests conducted in 1908 (to compare the British and U.S. built engines) give an interesting insight into the operation of freight trains on the steep climb up the escarpment from the coast. The usual load was six bogie wagons weighing between 155 and 160 tons with the locomotive worked from full gear to the third notch and regulator only part open. One can imagine these little engines must have barked as they stormed up the grade! The return journey was accomplished tender first and one can assume safely (after the reported problems with tenders) that riding the footplate instead of the tender was a safer proposition, even though riding of both engine and tender was probably pretty lively! It was normal to run the trains up the hill as a sort of caravan, one after the other, with the first leaving at 4.45am. The first couple of trains were only loaded to 100 tons or so and their job was to "to clear the rails" of a peculiar greasiness caused by heavy falls of dew which even sanding failed to overcome and appears to have been caused by the shut in nature of the line by the surrounding high hills. This caused slipping and for all intents and purposes was more of a problem than the steep grade and sharp curves. With the first trains having overcome this problem, later trains were built up to full load. It is also interesting to note reference in these reports to some of the British engines performing better than the U.S. engines with the former "catching up" to the latter as they ascended the grade!

Extensive coal tests were performed by David Buttle, the Locomotive Suprintendent in 1909, comparing Australian coal, Vipond coal, and U.K. Patent Fuel. As well as proving the worth of the Patent Fuel, the tests dramatically highlighted the superiority of the U.K. built locomotives over the American with regards to steaming. In tests which represented every day all-round work, the British locomotives on average showed a lighter fuel consumption of 19.53%, this superiority in steam more than offset the higher maintenanace costs and was the death knell for U.S. builders.

The 2ft 6in gauge railway which the FCAB operated should not be thought of in the context of European narrow gauge. As an example the 2−8−0s built by Hunslet and North British in

A panorama of the moles at Antofagasta. FCAB

1908–11 had 16½in × 20in cylinders, 37½in coupled wheels, 10ft 7in wheelbase, a heating surface of 1148 sq ft of tubes, 88 sq ft firebox and 20.1 sq ft grate area, with a boiler pressure of 180lbs. The tender carried 6¼ tons of coal and 2500 gallons of water with a weight for the engine and tender in working order of 74 tons 2 cwt for the Hunslets and 77 tons 1 cwt for the NBL locos. These were quite large engines for the gauge which was often overlooked because of their low chunky appearance.

The FCAB inherited from the Huanchaca Co. a regular International passenger train service to Bolivia. To operate this service the railway employed some very attractive American style, end platform clerestory roof corridor coaches, which were 49ft long, 7ft 8in wide and had a tare in the 17½ ton range. These coaches covered first and second classes, and there were also very comfortable (from contemporary reports) side corridor sleepers and diners of similar dimensions used in the trains. The goods stock in the main also followed American style and practice, with quite large gondolas and bogie box cars with brake wheel stanchions and foot boards along the roof. In stark contrast were a family of short, very British style four wheel open wagons, used primarily for mineral traffic, that would have looked equally at home on the Vale of Rheidol or Welshpool & Llanfair! In addition were some interesting six wheel flat wagons which were used for bagged nitrate shipments as well as other goods. Despite the American appearance of much of the rolling stock, a great deal of it was constructed by British rolling stock builders.

In May 1906 a contract was signed between the Bolivian Governmnent and the National City Bank and Speyer & Company, of New York to provide for the construction and operation of railway lines in Bolivia. The two companies then promoted the organisation of a separate entity to be known as the Bolivia Railway Company. It was incorporated on 18 February 1907 and had assigned to it the above concession. Construction had started on a 128 mile (207km) line connecting Oruro and Viacha in 1906—this being opened to traffic during October 1908. Agreement was reached on 19 October whereby the FCAB would lease the line from the Bolivia Railway Co., and from 1 January 1909 would take over its working. The FCAB constructed the railways in Bolivia in the capacity of a civil engineering contractor to the Bolivian Government but these lines could not be officially leased to the Bolivia Railway Co. until the Bolivian Government had satisfied the Bondholders from whom the money to construct the lines had been borrowed. The Oruro–Viacha section was leased and operated by the FCAB from 1 January 1909.

The branch lines in Bolivia were also financed by the Bolivian Government and then leased to the Bolivia Railway Co. who sub-contracted with the FCAB for their operation. A 56 mile (90km) branch from Uyuni to Atocha was begun in 1911, opened to traffic on 1 March 1913 and accepted under an FCAB operating lease from 1 July 1920. Work was inaugurated on the 108 mile (174km) branch from Rio Mulato to Potosi in 1909, being provisionally opened for services during 1912 and passing to the FCAB under an operating lease on 1 January 1916.

Work on the Viacha to La Paz section 20 miles (32km) to give access to "the highest capital in the world" was completed in 1917, under a concession granted in 1913. A further concession was then obtained and the line re-located, this work being completed in 1924.

Work started in 1909 on the 127 mile (205km) branch from Oruro to Cochabamba and this was completed by 26 July 1917. The FCAB operating lease was not activated until 1 January 1924.

On 5 January 1909, the FCAB floated a wholly owned subsidiary, the Compañia Ferrocarril de Aguas Blancas to acquire the Ferrocarril Caleto Coloso a Aguas Blancas 128 miles (206km) in length which extended from the small open roadstead port of Caleto Coloso six miles south of Antofagasta, running to Aguas Blancas with branches to the nitrate fields. The original company which owned this line started construction in 1898. In May 1911 a 10km long connecting link was laid in the vicinity of La Negra, on the FCAB main line, but for operational purposes the Aguas Blancas system was treated as a branch line. Locomotives retained by the FCAB following the takeover were numbered into a special 500 series block, as were other units specifically allocated to this branch.

The locos listed as taken over by the FCAB in January 1909 are from the FCAB's CME Return to the London Head Office. A couple of these engines are a mystery—the identity of the Shay is doubtful, and the builder of the little 0–4–2T No. 27 of 1907 is also unclear. Several enthusiasts have questioned Krupp as the builder at this early date, but the return referred to above quite clearly shows A. Krupp. One source has suggested that as the name of Krupp was becoming notorious for heavy armaments the clerk writing the list confused Koppel with Krupp. As the FCAB had never had Koppel engines, the shortened form of Arthur Koppel—AKop—may have been unknown to the clerk and easily extended as A. Krupp. The same return also credits the FCAB as the builder of No. 1. This is explained by the fact that this unit was an FCAB rebuild. Contemporary returns list it as having 37in coupled wheels, 15in × 18in cylinders, the same specifications as the 1895 Baldwins. The origins of this engine is not known. It is doubtful however, if it was one of the Huanchaca purchased Baldwins as 7 were transferred upon termination of the lease and the first, No. 46, was retained by Huanchaca and sighted as late as 1955.

Following the takeover of the Bolivian metre gauge lines, the FCAB formulated two separate self managing admistrations, each with a General Manager, CME, Civil Engineers, etc. One was based in Antofagasta to administer the 2ft 6in gauge lines and the other in La Paz for the metre gauge lines.

Work was also started in 1907 on a 69 mile (111km) branch from O'Higgins at km36 on the main line, to Boquete and the nitrate fields of Pissis and Domeyko.

As well as building new lines the company was also upgrading the existing main line to Uyuni and building new workshops. The 2ft 6in gauge was originally laid with 24ft long 36lb rails. 50lb rail was subsequently used and in 1907-8 the track up to and throughout the nitrate districts was relaid with 30ft long 65lb rail. Sleepers were mostly of Chilian oak 6ft × 8in × 5in while creosoted red pine and Quebracho had also been used. The Chilian oak sleepers weighed 26 kilos and the latter 66 kilos. It is worth noting that these 2ft 6in gauge sleepers were 6ft long compared to metre gauge sleepers which were only 6ft 6¾in long! In the desert areas of Chile the sandy earth acted as ballast but in areas subject to precipitation and in Bolivia, stone ballast was used. The railway has no tunnels, but earthworks are extensive, with stone culverts, and bridges, some of them quite large, are of steel construction.

Around 1910 the main Chilian workshops were transferred to new buildings at Mejillones. The engine shop as built was 540ft long and 220ft wide. The erecting bay was designed to accommodate 18 main line locomotives and fitted with two 35 ton overhead cranes. In adjoining bays were boiler, coppersmith and fittings shops, foundries for brass and iron, smithy and a machine shop. Next to this was a separate paint and carriage trimming shop 328ft long and 50ft wide. In line with this shop was the carriage and wagon shops, again 328ft long, but with two 50ft wide bays. The sawmill was situated in a separate 328ft by 50ft building feeding into the carriage and wagon shop.

A short distance from these workshops was a roundhouse with a capacity of 42 engines together with the usual offices, stores, coaling banks and water tower. Power for the complex was provided by two 110kva coal fired steam generating sets. Built at the same time were two slipways with a capacity of up to 280 tons, used for maintaining the company's maritime fleet of 6 steam tugs and over 120 lighters used at the ports of Antofagasta and Coloso.

The Uyuni works serving the Bolivian sector were smaller with two 290ft × 45ft bays and one 170ft × 45ft bay. The erecting shop had accommodation for 9 engines. Inherited from the Aguas Blancas railway was a workshop complex at Coloso built in the mid-1890s and housed in a building that was 410ft long and 82ft wide. Separate buildings housed carpenters and coach shops. There was a roundhouse with roads for 20 engines as well as all the auxiliary features an independent railway of 135 miles length would require.

This expansion was fueled by the nitrate boom. When the FCAB took control of its operations in 1904 only one nitrate oficina (or factory) was situated on the main line, though there were numerous sidings where untreated caliche—nitrate bearing earth—was loaded for shipment to Antofagasta for processing. With the discovery of the Shanks process of leaching and developments in Chili by J. T. Humberstone which made on site processing economically viable, the number of oficinas on the Antofagasta pampa alone had expanded to 20 by 1908. Where the nitrate deposits finished the copper belt started with the Chili Exploration Company mine at Chuquicamata, at an altitude of 8,846ft, while at the Bolivian border the Ollague branch fed in the output of the Collahuasi mines. Bolivia contributed borax, tin, silver and other concentrated ores to the freight carried. Nitrate however was the principal freight carried—in 1904 only 38,000 tons of manufactured nitrate was carried, but to this must be added 172,000 tons of caliche, whilst in 1908 no less then 379,000 tons of nitrate and no caliche was carried. With the expansion of down traffic in minerals there was a corresponding increase in up traffic particularly coal, which in 1904 amounted to 21,768 tons but had increased to 189,049 tons in 1908.

In 1912 an extension was opened by the FCAB from Cerro Negro to Augusta Victoria, a distance of 26 miles (42km), in the Boquete region.

Locomotive fuel at this time was coal, or Crown Patent Fuel,

The 'International' train used between Antofagasta and La Paz poses for its official portrait. The train consists of Baggage/RPO, Day Coach, three Sleeping Cars and a Restaurant/Dining Car. The cars, although they followed American practice, were actually British built by companies like Gloucester and Craven. The 2—8—4Ts were frequent performers on such trains.
FCAB

in the main shipped from U.K. ports, with spot shipments from the U.S.A., Australia and the southern Chilian coalfields, dependant on price and availability of shipping. In 1911 the FCAB placed an order for two 7,000 ton steamers with the intention of carrying fuel and stores outbound, and to backload with nitrate for the return voyage to the U.K. In 1912 however it was apparent that oil fuel from the expanding Mexican and Californian fields would be more cost effective, and the two steamers were sold whilst still on the stocks, showing a profit of £20,000. In the years immediately prior to World War I the company imported 120,000 tons of locomotive fuel per annum, its cost and shipping tariffs were closely monitored. In 1913 in line with the decisions made by many of the Nitrate oficinas, the company changed its fuel policy in favour of petroleum, that year saw a 250% increase in petroleum traffic over 1912.

Initially 22 locomotives were intended for conversion, but wartime shortages of steel boiler tubes saw only 8 completed by mid-November 1915. Full conversion of the roster only occurred after U.S. entry into the War in 1917, when supply problems in Chile eased. Post World War I coal fuel when used, was confined to the Chilian section of the main line, and coastal yard shunting engines. Freight locomotives were converted between fuels to take advantage of ruling prices, and in later years Government policy, which regulated the use of a percentage of native coal.

Freight movement was closely tied to commodity pricing and was subject to fluctuation. Taking 1925 as an example, at which stage the company operated 1,671 miles of railway, the Chilian section carried 399,311 passengers against 413,086 in 1924, while in Bolivia 264,476 were carried against 235,680 in the previous year. Goods traffic on the Chilian section amounted to 1,564,941 tons while on the Bolivian section 272,835 tons were carried. By this date the company had an extensive fleet of tank cars and was carrying the bulk of Bolivia's petroleum requirements.

From the late 1890s a number of small companies had commenced working copper deposits in the Chuquicamata region including Mr. Norman. A. Walker mentioned above. In 1910 Albert C. Burrage, a Boston entrepreneur consolidated the various concessions and in the following year sold off his Chilian interests to the Guggenheim Brothers, of New York.

Stock certificate for Compañia Salitrera Progreso de Antofagasta. The railway line on the map is probably a proposed railway by a partnership of Watson and Meiggs (of Peruvian railway fame) who secured a concession for a railway from Mejillones to Caracoles, a distance of 62 miles. Construction started 29 January 1873 and continued intermittantly until July 1875 when 35 miles of grading was complete along with a few miles of rail out of Mejillones. Most rail and equipment was lost in a tidal wave following an earthquake of May 1877. Rails and ties left intact were later recovered and sold at public auction, and were utilised to lay down the first street car line in Antofagasta. In later years a cart road followed this grading.

The four engines, of which No. 78 "Polpana" is a member, built by Hunslet in 1906 were the first 2−8−2s owned by the FCAB. They were built to the design of the FCAB's consulting engineers, Livesey, Son & Henderson and incorporated piston valves−an early application−with 15in×20in cylinders, 37½in driving wheels and weighed nearly 65 tons in working order. Two of the class, including this engine, were sold to the Boquete Nitrate Co. prior to 1912.
Hunslet Engine Co.

These Chilian purchases formed the basis of the Chilian Exploration Company (Chilex) founded in 1912. Actual construction of the Chuquicamata plant, and preliminary development of the pit, utilising the then revolutionary Jackling process, for exploiting low grade ores, with rail and electric shovel for working the mine started in 1913, and production at what rapidly became the worlds largest mine commenced in 1915. The mine subsequently passed into Anaconda control, but since nationalisation in 1952 has been worked by CODELCO the State Copper Authority.

To exploit the pit Chilex opted for a standard gauge network, in view of the tonnage expected. The system was named Ferrocarril Mineral de Chuquicamata, and opened in 1914. By 1928, the mine segment of this railway extended 67.4km and the plant segment, mainly third railed to take FCAB rolling stock contained 59km of line and connected with the FCAB branch which terminated 11.5km from the centre of the plant. In addition there was ½km of 30in gauge, plus sidings, with 500 3 ton cathode trucks to move copper from the electrolytic tank house to the furnace refinery.

The FCMC's return to the Chilian Government for the calendar year ended December 1928, shows a total of 9,743,663 short tons of ore and 5,699,643 tons of overburden removed from the mine, the works system accounted for a total of 8,981,209 short tons of tailings. The FCAB delivered 86,785 short tons of freight inwards, and shipped out 131,879 tons of refined copper. In addition the railway moved 4,000 workmen per day. To work this system in 1928 required 70 locomotives, both steam and electric plus some 700 wagons including 380 70 ton gondola ore cars, 80 20 yard air-dump waste cars, 120 12 yard hand dump tailing and waste cars, and 78 flats, of which 28 were equipped with iron sides for transport of workmen. There were 2 box-cars for transportation of explosives.

The only access to this enterprise, apart from a gravel road, was via the FCAB, which supplied all of its water requirements in addition to freight movements. All of the material to build the reduction works and refinery, housing for 4000 workers and their dependents, and the complete railway including the initial 37 locomotives, was hauled in on the 2ft 6in gauge over a 2 year period. At the time of construction the American locomotive industry was heavily committed supplying Allied requirements for the Great War. Locomotives were at a premium and the Guggenheims were fortunate in acquiring at least 23 (probably all 24) regauged Isthmian Canal Commission "100" Class locomotives disposed of in 1914 following completion of the Panama Canal. The history of Panama Canal locomotives is documented by Charles S. Small in his *Rails to the Diggings* (Railroad Monographs U.S.A. 1981). The "100" Class were built by Alco Schenectady in 1905 as 2−6−4T with 19in × 26in cylinders, weighing approximately 92 tons (ICC No's 101−124 Alco Works No's 38174−97). Too long for the rough construction tracks they were converted to 2−6−0 and fitted with sloping back tenders by the Panama Rail Road between 1907 and 1910. Unfortunately we were not able to unearth photographs of these engines been transported over FCAB metals to Chuquicamata.

Ongoing Chilex projects caused the FCAB to acquire special rolling stock, generally of U.S.A. manufacture, including bogie flat wagons of 45 ton capacity, and a massive 60 ton capacity well wagon, which could carry 20 tons of lead ballast to lower its centre of gravity it required. Carried on two 3 axle bogies it could be swung out of centre, for negotiating sharp curves via a ratchet arrangement. This car was used for carrying components for Chilex gyratory crushers.

In December 1924 the Guggenheims purchased the Anglo-Chilian Nitrate Company. The acquisition was part of a move, which within 10 years saw them with a virtual monopoly on the Chilian nitrate industry, including the takeover of the large British Lautaro company. Guggenheim's brilliant consultant metallurgist, and later senior employee E. A. Cappelton Smith, who had already introduced cost effective technology at Chuquicamata, developed a far more efficient form of nitrate reduction, which became known as the Guggenheim process. The now traditional Shanks process, introduced in the very early 1900's which equipped all of the FCAB serviced Oficinas on the Antofagasta and Boquete Pampas had an efficiency of between 55 and 75% in producing nitrate from Caliche, the raw nitrate rich earth. Guggenheims patented process had an efficiency of better than 90% thus making all of the FCAB serviced Oficinas obsolete, and by 1930 bankrupt, thus ending FCAB nitrate traffic.

In acquiring Anglo-Chilian, Guggenheims gained control of the F. C. Tocopilla al Toco, a 3ft 6in gauge nitrate network, built out of Tocopilla, an open roadstead port to the north of

Mejillones. Oficina Maria Elena, the southern terminus of this line, and the site of one of the giant new refineries lies only 90km due west of Chuquicamata who were of course the FCAB's largest corporate client. Only Chilian Government insistance on all new lines in the north of the Country being constructed to metre gauge, (and thus the conversion of the FCTT) delayed construction of a branch of the FCTT from Maria Elena to Chuquicamata. The project was finally killed by the Wall Street Crash. Guggenheims control of Anglo-Chilian resulted in all nitrate traffic shipped south on the Longitudinal line to Antofagasta, been transferred to Tocopilla. As a final blow Chuquicamata had been converting to full electric power, generated on site for both transport and processing. Chilex installed new electric power generating stations at Tocopilla and Maria Elena, thus denying the FCAB its existing fuel traffic.

LOCOMOTIVE LIST

1.4 Ferrocarril de Antofagasta a Bolivia

Original Numbers	1908 Numbers	Type	Builder	Works Numbers	Year Built	Remarks
75–78	109–112	2-8-2	HE	888–891	1906	75/78 sold to Boquete Nitrate Co. prior to 1912. 76–77 unserviceable in 1917 report.
1–3	12–14	0-6-4T	HE	878–880	1905	
4,5,7,8,10,11	15–20	0-6-4T	HE	907–912	1906	5/7/8/11 to Aguas Blancas Railway as 506–509 after 1912.
79	113	2-8-0	BLW	27995	1906	
80–81	114–115	2-8-0	BLW	28029–28030	1906	
82–83	116–117	2-8-0	BLW	28276–28277	1906	
84	118	2-8-0	BLW	28282	1906	
55–56	5–6	0-6-2T	HC	782–783	1906	
119–128	119–128	2-8-2	HL	2674–2683	1907	
129–138	129–138	2-8-2	HE	922–931	1907	
139–140	139–140	2-8-2	HC	787–788	1907	
45–50	45–50	2-8-0	Alco C	44617–44622	1908	
63–70	63–70	2-8-0	Alco C	44623–44630	1908	69 scr. in 1916 report.
85–90	85–90	2-8-0	Alco C	44631–44636	1908	
36	36	2-6+6-4T	K	4534	1908	Livesey, Son & Henderson type Meyer.
21–26	21–26	0-6-4T	HE	945–950	1907	
141–150	141–150	2-8-0	HE	958–967	1908	149 scrapped after accident La Cumbre 4/11/1913.

"Chuquicamata" is the name bestowed on 0-6-4T No. 7 built by Hunslet in 1906 as part of a batch of ten engines supplied in 1905 and 1906. No. 7 soon became No. 17 in the 1908 renumbering and later went to the Aguas Blancas Railway where it became No. 507. Many of the Aguas Blancas locomotives served for a lot longer than their contemporaries on the main FCAB, some lasting until the 1950s, and of course they remained 2ft 6in gauge.   Hunslet Engine Co.

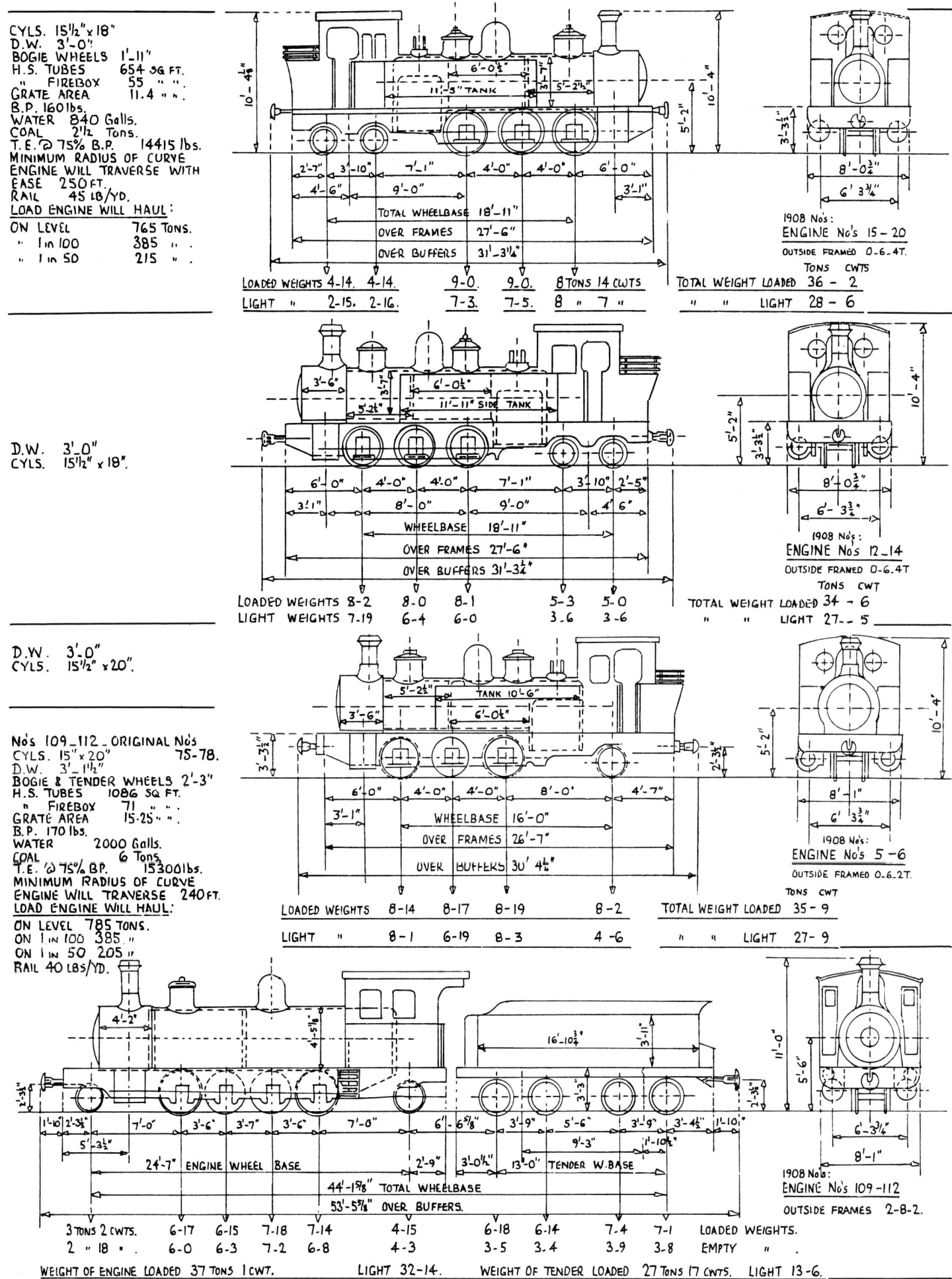

A double headed freight train stops for water on the 2ft 6in gauge FCAB main line c.1923–24. The lead engine is No. 115, a 2–8–0 built by Baldwin in 1906 as No. 81 and renumbered in 1908. Although the engine remains in substantially original condition it has been converted to oil firing, now has automatic couplers and is fitted with Westinghouse straight air brakes, the latter fitted in 1908.

J. M. Turner collection

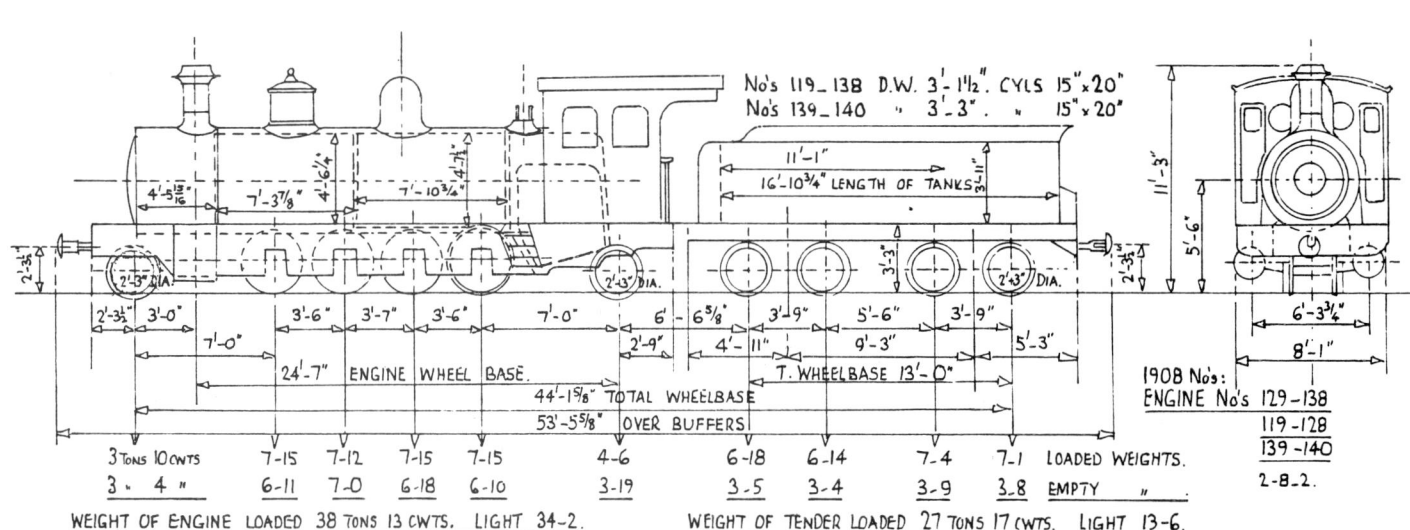

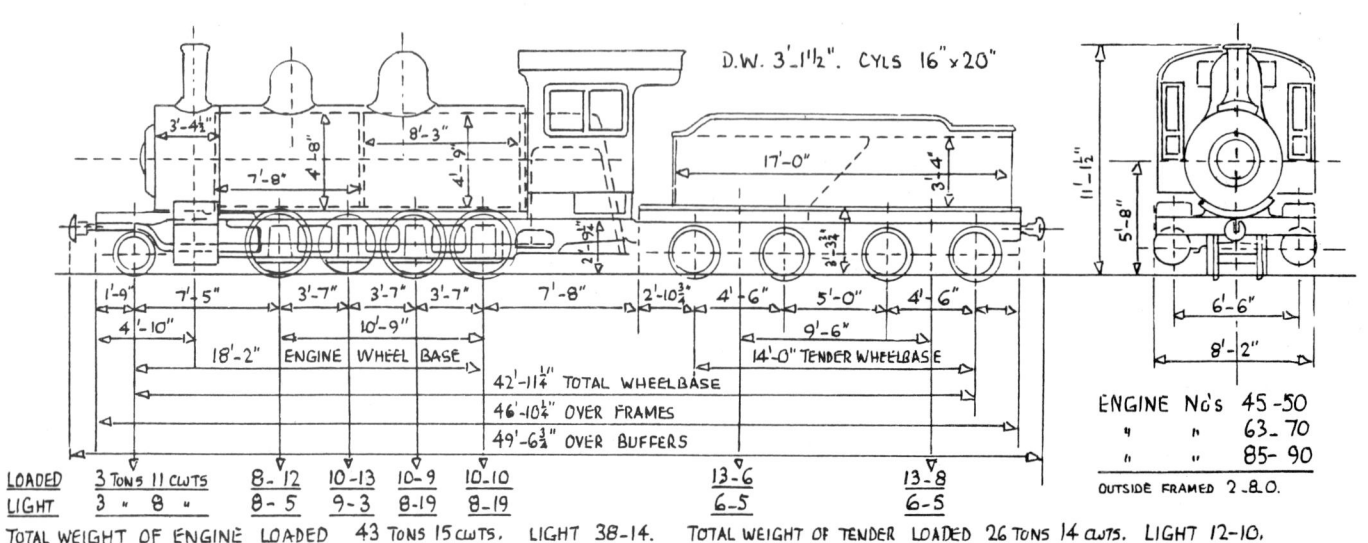

Alco 2–8–0 No. 45 "Curico" has only recently been completed and has been pulled out of the erecting shop at Schnectady in the winter snow to have the official works photo taken. The loco was one of a large order for 2–8–0s which were built by Alco in 1908. It looks as though it would be equally at home on the Sandy River line in Maine!   *Alco Historic Photos*

No. 25 was one of a group of six large 0–6–4Ts built to the 2ft 6in gauge by Hunslet in 1907 and had 15in × 18in cylinders and 36in driving wheels. This was a successful design by Hunslet which was also built in metre gauge for contractors in Chile and Brazil and in 3ft gauge for Russia.   *Hunslet Engine Co.*

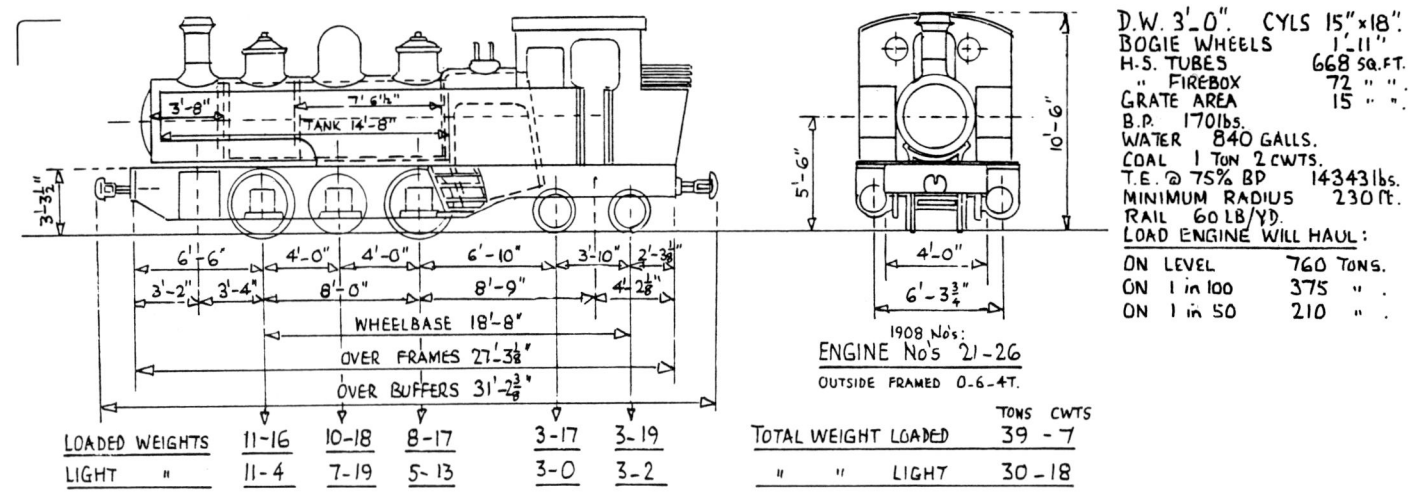

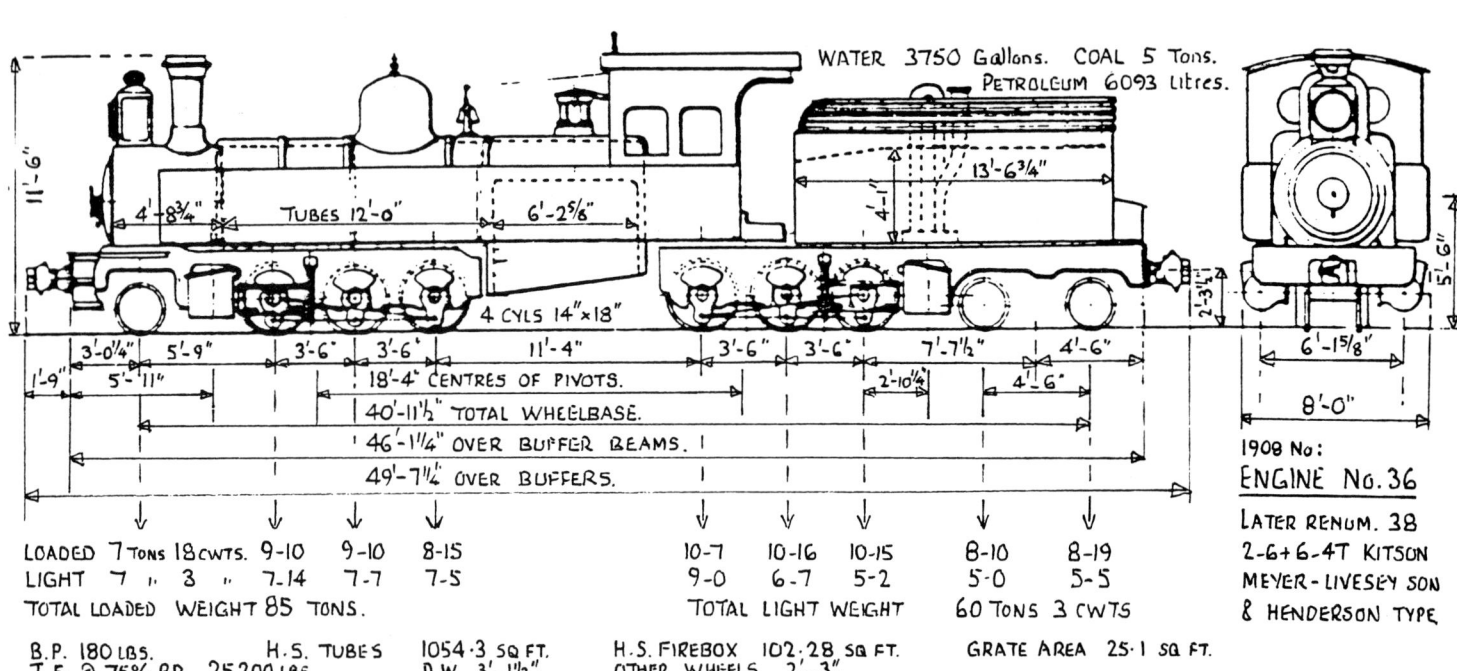

Livesey, Son & Henderson Meyer type 2-6+6-4T built by Kitson & Co., in 1908. Originally numbered 36 it was later renumbered 38. This version of the Kitson Meyer was not duplicated on the FCAB although a similar machine was built for the Leopoldina Railway in Brazil. D. Binns collection

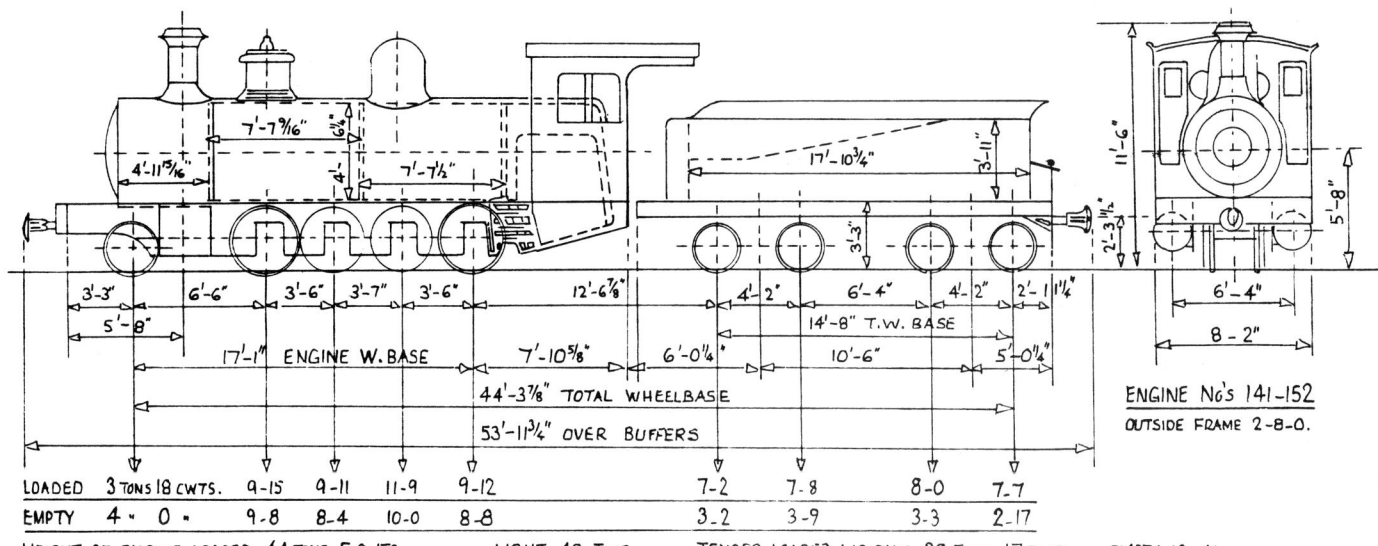

ENGINE No'S 141-152
OUTSIDE FRAME 2-8-0

	LOADED								
LOADED	3 TONS 18 CWTS.	9-15	9-11	11-9	9-12	7-2	7-8	8-0	7-7
EMPTY	4 " 0 "	9-8	8-4	10-0	8-8	3-2	3-9	3-3	2-17

WEIGHT OF ENGINE LOADED 44 TONS 5 CWTS.   LIGHT 40 TONS.   TENDER LOADED WEIGHT 29 TONS 17 CWTS.   EMPTY 12-11.

```
CYLS.          16½" x 20"
D.W.           3'-1½"
BOGIE & TENDER WHEELS  2'-3"
H.S. TUBES     1148 SQ FT.
 "  FIREBOX    88  "  "
GRATE AREA     20  "  "
B.P.           180 LBS.
WATER          2500 GALLONS
COAL           6¼ TONS
T.E. @ 75% BP  19600 LBS.
MINIMUM RADIUS CURVE   240 FT.
RAIL           60 LBS/YD.
LOAD ENGINE WILL HAUL:
ON LEVEL       1015 TONS
 "  1 IN 100   495  "
 "  1 IN 50    270  "
```

```
No. 146 ARRIVED ANTOFAGASTA  20-7-1908
LANDED               23-7-1908
ERECTION COMMENCED   30-7-1908
UNDER STEAM          13-8-1908
TRIAL RUN TO NORTH YARD    14.8.1908
AND TO PORTEZUELO          15.8.1908
IN SERVICE 26.8.1908.
```

Side and front works photos of No. 146 "Diana", a 2-8-0 built as part of an order for ten 2ft 6in gauge locos from Hunslet in 1908. The firebox behind the rear driving wheels gave these engines a rather lop-sided appearance and probably a tendency to hunt at any sort of speed. The front pony truck inside the frames must also have presented problems on sharp curves. The front view of the loco highlights the extreme width of FCAB narrow gauge locos.
Hunslet Engine Co.

1.5 Ferrocarril Caleta Coloso a Aguas Blancas

LOCOMOTIVE LIST

Formed in 1898 and opened in 1902 to service the nitrate pampa south of the FCAB main line. Taken over the Compañìa Ferrocarril de Aguas Blancas (which was wholly owned by the FCAB), in 1909, and thereafter was operated as a branch of the FCAB, connection having been made with the FCAB soon after 1909. Engines allocated to the line were renumbered in the 500 series, probably when they were taken into FCAB stock on 24/3/1909.

Original Numbers	FCAB Numbers	Type	Builder	Works Numbers	Year Built	Remarks
1	513	0–6–2ST	FCAB	–	1902	FCAB rebuild, probably of a Baldwin.
2–3	514–515	2–8–0	Rogers	5701–5702	1902	
4	524	2–8–0	BLW	24444	1904	
5	503	0–6–0T	Hen	6489	1903	
6	504	0–6–0T	Hen	7491	1906	
7	–	0–6–0T	Hen	6490	1903	Sold in 1911.
8–9	511–512	0–6–2ST	Rogers	6270–6271	1905	
10	516	2–8–0	Alco R	38445	1905	
11	519	2–8–0	Hen	7551	1906	
12	520	2–8–0	Hen	7753	1906	
13–14	517–518	2–8–0	Alco R	41115–41116	1906	
15	521	2–8–0	Hen	7754	1906	
16	–	0–6–2T	Hen	7550	1906	
17	–	0–6–0T	Hen	7549	1906	
18–19	–	0–6–2T	Hen	7958–7959	1907	
20	–	0–6–0T	Hen	7960	1907	Sold in 1911.
21	501	0–4–0T	Hen	7995	1907	
22	–	0–4+4–0TG	Lima	1677	1906	Shay—Not positively identified, only shown in lists as Shay, number, name and built 1907. This loco is probably 2ft 6in gauge No. 4 built for American Smelters Securities Co., Santa Barbara, Chile, desp. 12/4/1906 and is the only Shay in Koch's list around this date. Possibly bought or taken over in 1907.
23–24	522–523	2–8–0	Hen	8355–8356	1908	
25–26	–	0–6–2T	Hen	8353–8354	1908	Originally ordered by Compañìa Salitrera Nueva Castilla.
27	–	0–4–2T	?		1907	Not positively identified; supplied through the agency of A. Krupp.

0–6–2ST No. 8 "Iquique" was built by Rogers in 1905 as one of a pair for the 2ft 6in gauge Aguas Blancas Railway and was purchased through the agency of W. R. Grace. These engines were used for shunting and short trip working on this line which served a mining area to the south of Antofagasta.
Alco Historic Photos/Ron Redman collection

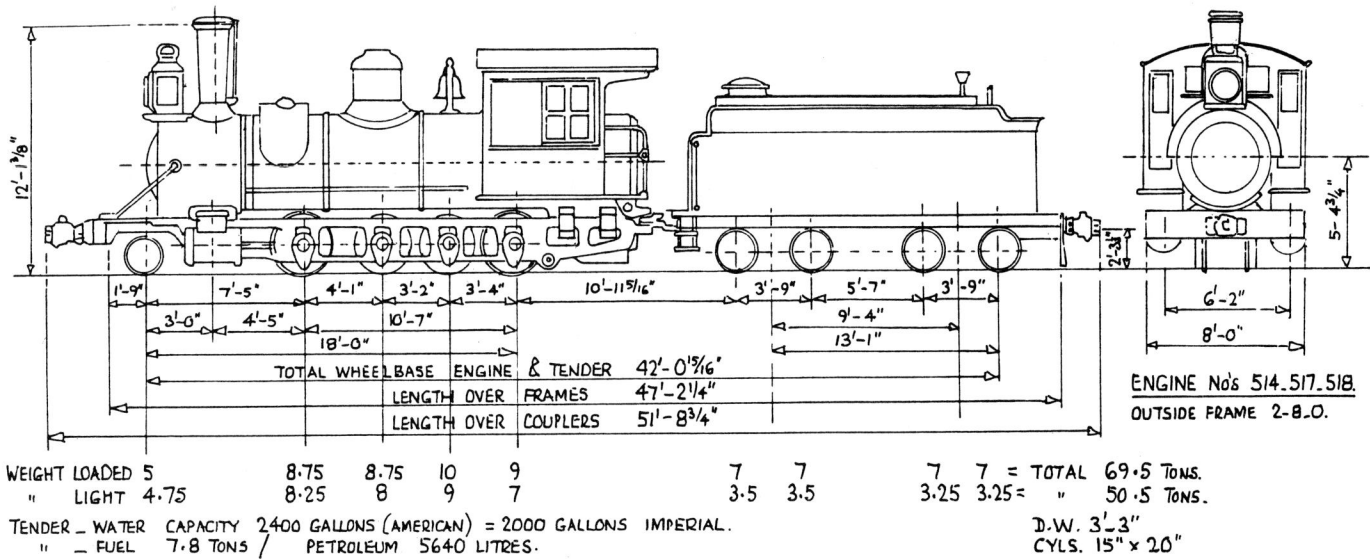

ENGINE No's 514.517.518.
OUTSIDE FRAME 2-8-0.

WEIGHT LOADED 5 8.75 8.75 10 9   7   7   7   7 = TOTAL 69.5 TONS.
"   LIGHT 4.75 8.25 8   9   7   3.5 3.5 3.25 3.25 = "   50.5 TONS.
TENDER – WATER CAPACITY 2400 GALLONS (AMERICAN) = 2000 GALLONS IMPERIAL.   D.W. 3'-3"
"   – FUEL 7.8 TONS / PETROLEUM 5640 LITRES.   CYLS. 15" × 20"

2. **Antofagasta (Chili) and Bolivia Railway Co. Ltd. Post 1908 numbers for 2ft 6in gauge locos.**

This list also includes the 500 series locos allocated to the Aguas Blancas Railway. Locos built after 1908 did not carry names, though it is likely some earlier locos continued to carry their names after 1908.

M = indicates that the loco was converted to metre gauge in 1926–1928, or earlier if a date is quoted.

1908 Number		Previous Number	Type	Builder	Works Numbers	Year Built	Remarks
1		15	0-6-2T	AE	1195	1877	
2		17	0-6-2T	SS	3032	1882	Rebuilt from 2-6-2T prior to 1912.
3		36	0-6-2ST	BLW	9770	1889	Disposed of prior to 1912.
4		47	0-6-2ST	BLW	10995	1890	Disposed of prior to 1912.
5–6	M	55–56	0-6-2T	HC	782–783	1906	
7		51	0-6-2ST	BLW	12752	1892	
8–9		52–53	0-6-2ST	BLW	12753–12754	1892	Disposed of prior to 1912.
10–11		56–57	0-6-2ST	BLW	14220–14221	1895	
12–14	M	1–3	0-6-4T	HE	878–880	1905	No. 12 converted to 0-6-0T – see footnote.
15	M	4	0-6-4T	HE	907	1906	see footnote.
16–19		5, 7, 8, 10	0-6-4T	HE	908–911	1906	Sold to Aguas Blancas Railway after 1908 as Nos. 506–509.
20	M	11	0-6-4T	HE	912	1906	see footnote.
21–26	M	–	0-6-4T	HE	945–950	1907	see footnote.
27–32	M	–	2-8-4T	K	4843–4848	1911	Refer note section 1:2 0-4-4T Rogers
33	M	20	2-4-2	BLW	8215	1886	Converted to metre gauge 1909–1914.
34–35		41–42	2-4-2	BLW	10943–10944	1890	Disposed of prior to 1912.
36		–	2-6+6-4T	K	4534	1908	Renumbered 38 in 1914 to make way for metre gauge 4-6-2s. Livesey, Son & Henderson Meyer type. Scr. in 1929.
37		–	2-6+6-2T	K	4841	1912	Kitson Meyer type. Out of service by 1928.
42		21	2-4-2	Rogers	3713	1887	Disposed of prior to 1912.
43–44		61–62	4-4-0	BLW	14464–14465	1895	Scr. in 1916 report.
45–50		–	2-8-0	Alco C	44617–44622	1908	
51–62		–	4-6-0	RS	?	?	Remaining units from 1876 and 1887–1888 batches, individual numbers unknown. Disposed of prior to 1912.
63–70		–	2-8-0	Alco C	44623–44630	1908	69 scr. in 1916 report.
71–72		30–31	2-6-0TT	BLW	9846/9852	1889	71 scr. in 1916 report.
73–74		43–44	2-6-0TT	BLW	10984/10988	1890	
75		45	2-6-0TT	BLW	10997	1890	Scr. in 1917 report.
76–77		39/38	2-6-0TT	BLW	10464/10470	1889	
78–79		48–49	2-6-0TT	BLW	11426/11436	1890	
80		50	4-6-0TT	BLW	11437	1890	
81		37	2-6-0TT	BLW	10469	1889	Scr. in 1916 report.
82–83		32–33	2-6-0TT	BLW	9855/9864	1889	83 scr. in 1917 report.
84	M	34	2-6-0TT	BLW	9859	1889	Converted to metre gauge 1917. Out of service by 1925.
85–90		–	2-8-0	Alco C	44631–44636	1908	

1908 Number	Previous Number	Type	Builder	Works Numbers	Year Built	Remarks
91	63	4-8-0	Cail	2466	1898	Scr. in 1917 report.
92-94	58-60	2-8-0	BLW	14461-14463	1895	92 Unserviceable 1917 report
95	35	2-8-0	BLW	9773	1889	Scr. in 1916 report.
96	53 (2nd)	2-8-0	BLW	12635	1892	Scr. in 1917 report.
97-98	54-55	2-8-0	BLW	12633/12667	1892	
99-100	64-65	2-8-0	BLW	17461-17462	1900	100 scr. in 1917 report.
101-104	67-70	2-8-0	BLW	18388-18391	1900	101 scr. in 1917 report.
105-108	71-74	2-8-0	BLW	19437-19440	1901	
109-112	75-78	2-8-2	HE	888-891	1906	109/112 sold to Boquete Nitrate Co. prior to 1912. 110-111 unserviceable in 1917 report.
113	79	2-8-0	BLW	27995	1906	
114-115	80-81	2-8-0	BLW	28029-28030	1906	
116-117	82-83	2-8-0	BLW	28276-28277	1906	
118	84	2-8-0	BLW	28282	1906	
119-128	—	2-8-2	HL	2674-2683	1907	Rebuilt to 2-8-0s in 1918-1922. Five locos rebuilt and renumbered 525-9 after 1928. Remainder withdrawn by 1928.
129-138	—	2-8-2	HE	922-931	1907	All converted to 2-8-0s in 1918-1922. One loco rebuilt and renumbered 530 after 1928. Remainder withdrawn by 1928.
139-140 M	—	2-8-2	HC	787-788	1907	Rebuilt as 2-8-0s in 1918-1922.
141-150 M	—	2-8-0	HE	958-967	1908	149 scr. after accident 4/11/1913.
151-152 M	—	2-8-0	HE	1066-1067	1911	
153-160 M	—	2-8-0	NBL	19428-19435	1911	
161-170 M	—	2-8-2	HL	2943-2952	1912	168 converted to 2-8-2T in 1922 and renumbered 1.
171-180 M	—	2-8-2	Hen	11891-11900	1913	
501	21	0-4-0T	Hen	7995	1907	
502	renumbered from 504 below					
503	5	0-6-0T	Hen	6489	1903	
504	6	0-6-0T	Hen	7491	1906	renumbered 502
505	see note below					
506-509	—	0-6-4T	HE	908-911	1907	Ex No's 16-19
510	see note below					
511-512	8-9	0-6-2ST	Rogers	6270-6271	1905	
513	1	0-6-2ST	FCAB	—	1902	
514-515	2-3	2-8-0	Rogers	5701-5702	1902	
516	10	2-8-0	Alco R	38445	1905	
517-518	13-14	2-8-0	Alco R	41115-41116	1906	
519	11	2-8-0	Hen	7551	1906	
520-521	12/15	2-8-0	Hen	7753-7754	1906	
522-523	23-24	2-8-0	Hen	8355-8356	1908	
524	4	2-8-0	BLW	24444	1904	
525-529	—	2-8-0	HL	(2674-2683)	1907	Ex 119-128 series
530	—	2-8-0	HE	(922-931)	1901	Ex 129-138 series

ENGINE No's 27-32
OUTSIDE FRAMED 2-8-4T.

D.W. 3'-1½". CYLS 17"×22"

At an unknown date there was a major renumbering of 500 series locos to consolidate this group of numbers after some locos had been scrapped. Complete details are known only for 504 to 502, and 525–529 to 519–523. It is likely that the main line Hawthorn Leslie and Hunslet locos were stored in 1928 and later cannibalised to keep the Aguas Blancas 2–8–0s going. By this time both the Rogers 2–8–0s and Henschel 2–8–0s had been withdrawn.

Hunslet 0–6–4Ts: After 1908, but prior to the 1920s, a number of these were converted to 0–6–2Ts and 0–6–0Ts. There is an official photograph taken at the FCAB Shops showing No. 12 extensively rebuilt as an 0–6–0T, and later loco lists record at least six 0–6–2Ts, but details of which were converted, (other than these mentioned) to what wheel arrangement, and at what date have not yet been established from available records.

2–8–4T No. 27 was one of a batch of six engines built in 1911 with 17in×22in cylinders and 37½in driving wheels. They weighed just over 68 tons and were a standardised design with the Hawthorn Leslie 2–8–2s delivered the following year, sharing boiler, wheels, cylinders and motion. They all were converted to metre gauge and some lasted into the 1960s.

FCAB

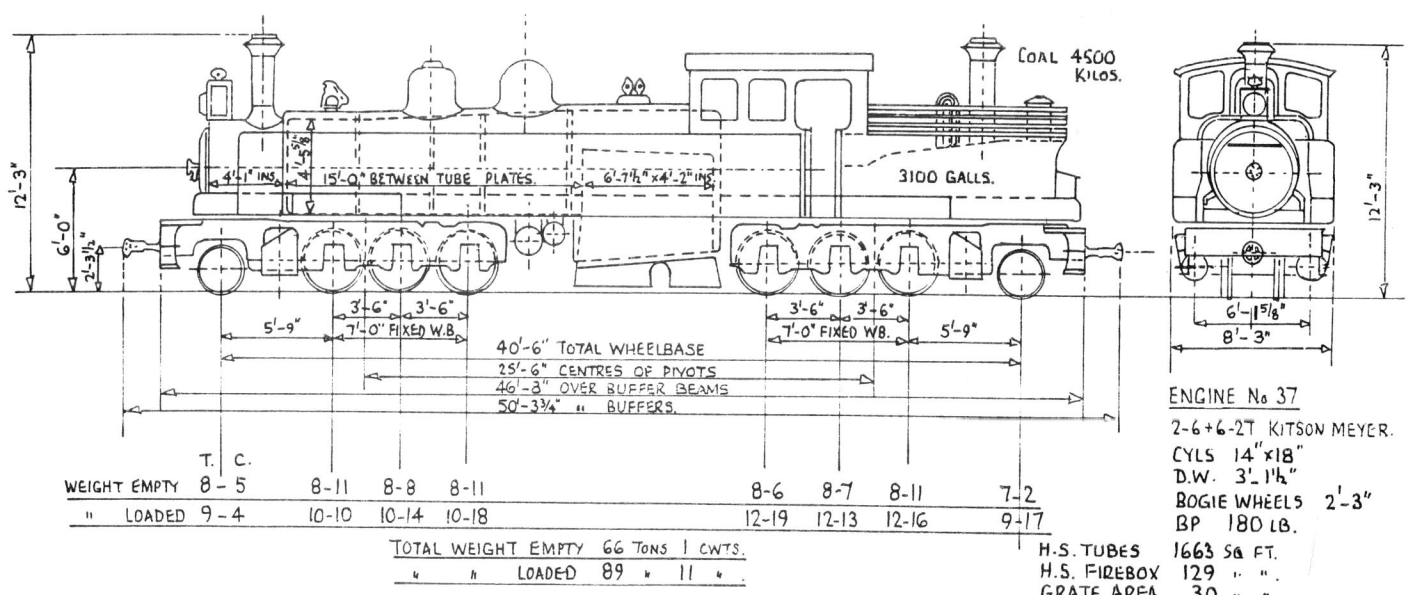

No. 37 was a conventional Kitson Meyer type built by Kitson & Co. in 1912. Due to problems in changing the gauge on this 2–6+6–2T it was out of service by 1928.

D. Binns collection

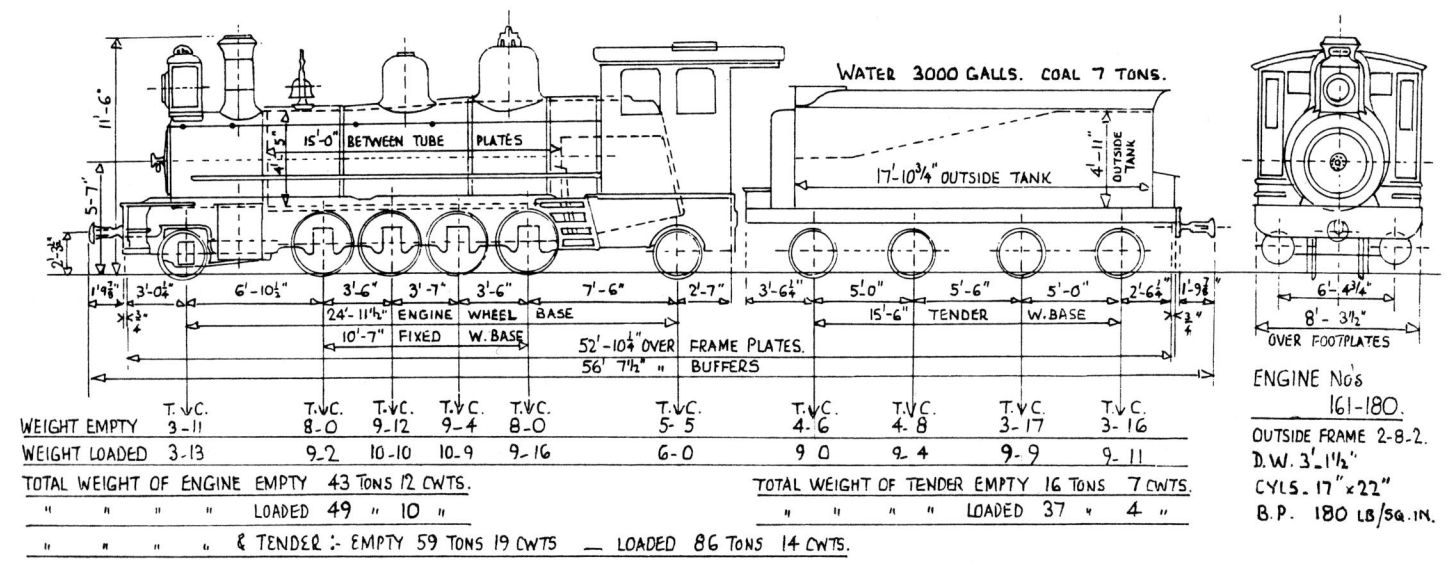

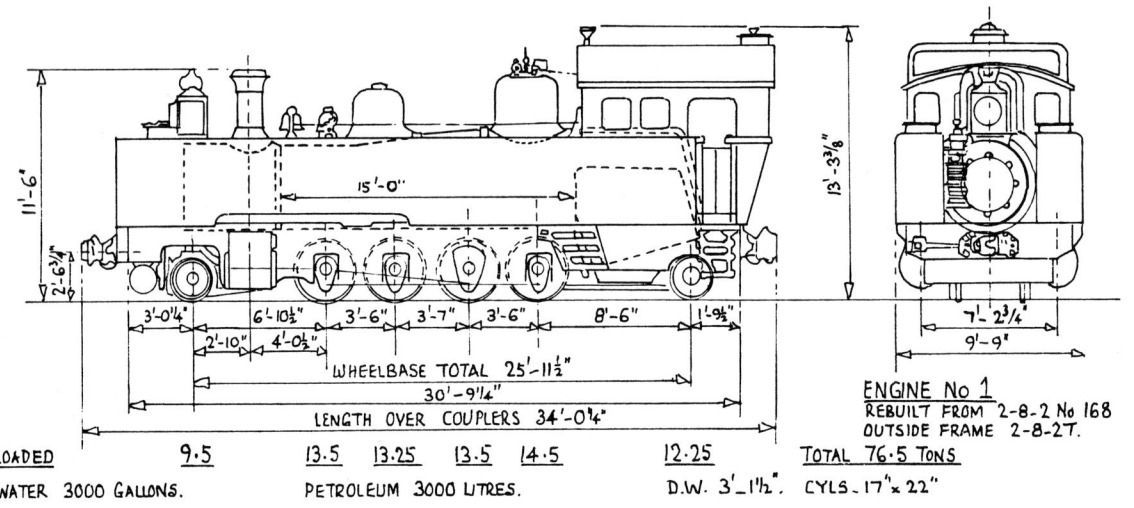

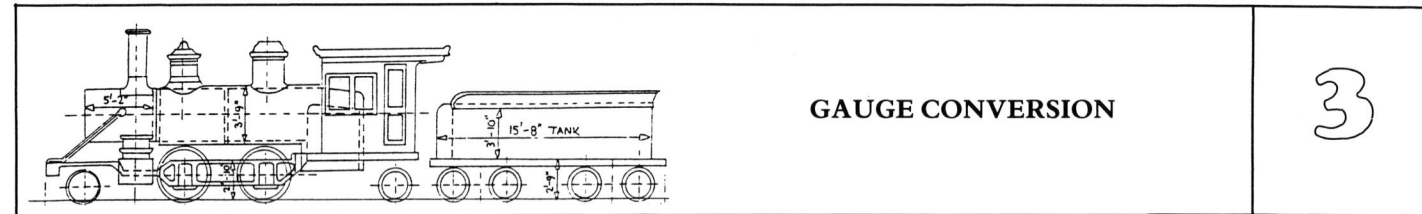

GAUGE CONVERSION 3

The development of international traffic experienced in the years immediately preceeding the First World War resulted in a Board decision in 1913 to convert the whole system to metre gauge. There were other considerations such as the construction to metre gauge of the Chilian Northern Longitudinal Railway, which crossed the FCAB main line at Baquedano mp60 (km97) as well as the policy adopted by the Chilian Government that metre gauge should, as far as practicable, be the standard gauge for all railways in Northern Chile. The lines of the Bolivian Railway Company had of course been constructed to metre leaving the FCAB main and branch lines in Chile at 2ft 6in gauge.

A fact not commonly known was that the FCAB was a pioneer in the bogie exchange of rolling stock between 2ft 6in and metre gauges and for through goods traffic to and from points north of Uyuni, a bogie changing apparatus was installed at Uyuni, by means of which the bodies of the wagons were lifted by jacks and the bogies changed according to the gauge of the track on which they were to run. The FCAB preferred this method of gauge interchange because it was less labour intensive than physically transferring loads.

World War One intervened and only sufficient material was available to widen the Oruro to Uyuni section which was completed in February 1916. About 1919 a third rail of metre gauge was laid on the main line from Antofagasta to Baquedano, along with part of the yards at Antofagasta. The reason for constructing this third rail was to provide through connection with the Chilian Northern Longitudinal Railway.

At the time of the third railing an unknown number of Longitudinal metre gauge locomotives were modified to incorporate offset couplers, in addition to their standard ones to enable them to handle 2ft 6in gauge rolling stock over the steep escarpment section from Antofagasta to Prat.

Although the decision to convert to metre gauge was taken in 1913 it would appear that the 2ft 6in gauge locomotives supplied by British builders and Henschel, starting with

Hunslets 1908 delivery, were built with eventual metre gauge conversion in mind. We have confirmed that at least two 2ft 6in gauge Baldwins were converted to metre gauge in this 1909–17 period, however, we believe others were converted in the 1909–11 period for construction work on the Bolivian branches but to date have not located any records listing specific locomotives.

In July 1919 the FCAB took over operations of the metre gauge Northern Longitudinal Railway, including the Chanaral Railway. The Longitudinal Railway consisted of a 441 mile north-south link from Pueblo Hundido on the Chanaral Railway in the south to Pintados in the north where a connection was made with the standard gauge Nitrate Railways. The Chanaral Railway is a metre gauge east-west system 168 miles long connecting the port of the same name with gold and copper mines in the Andean foothills. It was purchased by the Chilian Government in 1888. In turn the Chanaral Railway connects with the Red Norte, the State metre gauge northern network extending 943 miles south to La Calera near Valparaiso and a connection with the 5ft 6in gauge southern network. The FCAB operated the Longitudinal Railway on behalf of the Chilian Northern Railway Co. Ltd., a direct subsidiary, which in turn operated the line for the State.

As we have indicated, World War I intervened with plans to convert the Antofagasta to Uyuni main line to metre gauge and it was 1928 before this was finally undertaken.

By 1926, it was evident that, if the work was to be carried out, it should be done promptly. Some of the locomotives and wagons were approaching the end of their useful life, and would soon require to be replaced, and a large amount of track and switch material in branch lines and yards was due for renewal and needed replacing with a rail of heavier section. These and other considerations, as well as the advantages to be derived from the greater speed and carrying-capacity attainable on the broader gauge, decided the Board to authorise the carrying out of the entire scheme as soon as possible.

Third railing had progressed as far as Calama (mp 149) by June 1928 and some easy sections in Bolivia, where adding a third rail was practical, had also been completed.

The following paragraphs are extracted from *Facts and Views concerning the change of Gauge from 2ft 6ins to One metre* (FCAB June 1929).

"In spite of some delay in obtaining delivery of part of the material required, work proceeded rapidly, until by the end of June 1928, all was ready for the final stage of changing the gauge of the main line, and the necessary authority was obtained from the Chilian and Bolivian Governments to close the section between Calama (kilometre 239) and Uyuni (kilometre 617) for the period July 5th to 10th inclusive.

Of the 378 kilometres of line between Calama and Uyuni, 92 already had the third rail laid to metre gauge, leaving 286 kilometres of track on which the rails had to be spread, of which 173 kilometres corresponded to the Bolivian Section and 113 to the Chilian. This work had to be completed within six days, as stated above. To effect the change of some 48 kilometres of line per day is, in itself, no mean achievement, but the difficulties in the present instance were acentuated from various causes. It was mid-winter; the line is situated at altitudes varying from 10,000 to 13,000 feet above sea-level, and passes through a district very sparsely inhabited; suitable labour was scarce, and provision had to be made for housing the men engaged on the work and to supply them with their requirements of food and water.

On the Bolivian Section, that is, between Ollagüe and Uyuni, no three-railing was done, as, owing to the very large number of bridges, both rails had to be moved outwards, in order to preserve the alignment. Preparations were made by placing new coach screws to the required gauge and adzing the sleepers where necessary to give level seating for the rails in their new position. It was decided that 35 gangs of men would be required to complete the work in the stipulated time; the 35 camps were duly installed, meat, water and provisions being supplied for ten days, and two days before work was actually to be started, special trains were run from various parts of the

This view shows the arrival of the first metre gauge train, a passenger (and probably the International), at Ascotan at 11.15a.m. on 11 July 1928. The train is hauled by one of the converted 2ft 6in gauge 2–8–0s. At this time the track in the yard and on the main line was still dual gauge but the inside narrow gauge track would progressively disappear.
J. M. Turner collection

A typical scene on any single line railway, this time on the 2ft 6in gauge FCAB. Train crews confer as a double headed freight, with Hawthorn Leslie 2−8−2 No. 128 in the lead, waits to cross the International Train which has just arrived on the left hand track, c1927, at Ollagüe. On the right is one of the FCAB standard box cars (many of which were British built to American design) with its unusual handrail arrangement.
J. M. Turner collection

Left hand block of 4 pictures below − Top left: Diamond crossing North Yard, Antofagasta before alteration. Top right: After alteration. Bottom left: Special Switch, type 4. Bottom right: Special Switch, type 3A.

Republic bringing the men required. The greatest enthusiasm was shown, and although a late start was made on July 5th, the work was practically completed by midday on July 9th. A pilot-train was run on the same day from Uyuni to Ollagüe, and returned on the 10th July, picking up the men who had completely finished their sections.

An interesting feature so far as the Bolivian Section is concerned is that good weather prevailed during the whole of the time the work lasted, but on July 11th, when the first metre-gauge train ran from Bolivia to Chili, a very severe sand storm and heavy gales were experienced, these lasting for several days. Had such weather prevailed during the execution of the work, it would have been impossible to proceed, as the gales were so severe that trains were held up, the line, in many places, being covered with sand.

On the Chilian Section there is only one bridge of importance, but, on the other hand, there were a considerable number of stations to be dealt with, some of them important traffic centres provided with a considerable number of sidings, notably the frontier station at Ollagüe, and the summit station at Ascotan (13,000 feet above sea-level).

All difficulties, however, were successfully surmounted, the work was finished by July 9th; on the same day a pilot-engine ran over the new gauge; and on the evening of the 10th the first metre-gauge international train left Antofagasta on the run of 1,173 kilometres to La Paz, accomplishing the journey without incident."

The Boquete branch was opened to metre gauge working on 7 December 1926 and on the Mejillones branch mixed gauge working was instituted on 14 April 1927. The important Ollague branch to Collahuasi was opened on the wider gauge on 27 August 1928. The Chuquicamta branch was widened by spreading the track and was completed in October 1928.

Although the main line between the coast and Bolivia was opened to metre gauge on 10 July 1928, and a direct service of goods and passenger trains was established between Antofagasta

This historic view shows the last 2ft 6in gauge train arriving at Ascotan on 5 July 1928 thus bringing to an end some fifty years of narrow gauge main line operations, although some branch and industrial lines, including the Agua Blancas system remained 2ft 6in. The train of mainly oil tank wagons is hauled by Nos. 70 and 87 both built by Alco Schnectady in 1908 and interestingly from the same batch of locos. Whether this was by design or pure coincidence is unknown.
J. M. Turner collection

and stations to the north of Calama, there still remained a large amount of work to be done on the Chilian Section to convert the numerous branch lines, nitrate "Oficinas", and mining establishments, in addition to the heavy and complicated work of converting the tracks, moles, etc., in the ports of Antofagasta and Mejillones. This latter work had to be carried out in such a manner that the track conversions effected conformed to the constantly varying traffic requirements of the two gauges.

Complete metre gauge working came into operation on 6 December 1928, nearly two months before the scheduled time.

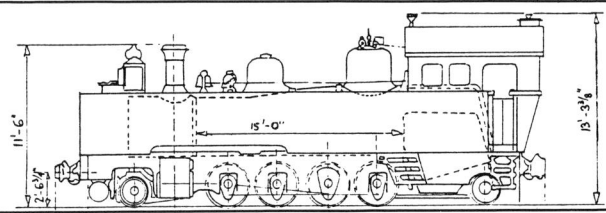

METRE GAUGE LOCOMOTIVES 4

The gauge conversion naturally entailed the ordering of new locomotives and rolling stock, however, due to the oversize dimensions of the 2ft 6in gauge rolling stock, and as there was little non bogie stock in service, conversion of the older stock at most only entailed a permanent bogie change, which was accomplished by converting some bogies or building new ones depending on suitability. Starting in 1908 Westinghouse straight air brake equipment had been fitted as had automatic couplers. As a result of the interchanging of bogies at Uyuni, as a means of goods transfer, rolling stock of both gauges were fully compatable. Orders for new motive power, rolling stock and conversion material were placed prior to 1926 and in July of that year a start was made on conversion. At the commencement of the programme, one or two locomotives and 60 to 80 wagons were processed each month, and in the same period two new locomotives and 25 to 30 wagons were erected. Towards the end of 1928 the rate of conversion for wagons had increased to between 250 and 300 per month. This entailed the dismantling and re-erection of from 500 to 600 bogies as well as the removal and refitting of new axles on about 1,000 to 1,200 pairs of wheels. Although the workshops facilities at the Company's disposal was installed for maintenance purposes only and were not intended to deal with such an unusual task, it is remarkable that in total the conversion programme handled some 61 locomotives, 103 carriages and 2,140 wagons in addition to its normal work.

The arrival of the newly built locomotives included some interesting designs. The metre gauge North British 2−8−4Ts were designed to haul double the load of the 2ft 6in gauge eight coupled locomotives. These tank engines had a rigid and total wheelbase of 11ft 9in and 31ft respectively, 19in × 24in cylinders, 44in driving wheels, boiler pressure of 180lbs and a tractive effort at 75% boiler pressure of 26,580lbs. Their tank capacity was 2,530 gallons of water and 1,478 of fuel oil, most of the latter being in a bunker incorporated into the cab roof! A feature of these locos not generally known was that they were adaptable to either coal or oil fuel and some did run as coal burners on occasion. The boiler, cylinders, motion and driving wheels of this class were of a standard design also used in the later 2−8−2s of the 900 series. In 1929 Hawthorn Leslie built three 4−8−2Ts for passenger service which in outline were similar to the North British 2−8−4Ts, again with oil bunker incorporated in the cab roof and with everything but the proverbial kitchen sink hanging on the outside totally obscuring the boiler! These three locomotives shared the standard boiler, however had 21in × 24in cylinders and 51in driving wheels, with a tractive effort of 31,752lbs at 85% boiler pressure. Though looking smaller than they actually are, neither of these two classes would fit within the British loading gauge and in working order tipped the scales at just over 86 tons!

The need for further articulated locomotives led to introduction in 1929 of three large 4−8−2+2−8−4T Beyer Garratts which weighed in at no less than 170 tons making them the largest FCAB machines to date. With a 13 ton axle load, they were fitted with bar frames, Walschaerts valve gear, Belpaire fireboxes and were equipped for oil burning. Their intended use was on the Potosi branch although there are photos in existence showing them working the heavy international passenger train out of La Paz (which may have been for test purposes however) and freight trains on the main line in Bolivia.

It is interesting at this time to review the locomotive fleet as it existed and grew in the period from the gauge conversion to the beginning of the Second World War. The FCAB started out with its converted British built 2−8−0s which must have looked decidedly small alongside the locomotives built for metre gauge! In addition were metre gauge conversions of the tank engine family of 0−6−4Ts, 0−6−0Ts and 2−8−4Ts. The Bolivian system added a series of Alco 2−8−0s whilst the Chilian Northern offered some FCAB design 0−6−4Ts, a family of Henschel 2−6−0s and a group of standard design 2−8−2s which came from North British and Beyer Peacock. The FCAB itself, in addition to the standard design 2−8−4Ts and 2−8−2s (and their 4−8−2T equivalent) also provided some Henschel 2−8−0s and very unusually, a group of Henschel 4−6−2s! The exact purpose of these Pacifics on such a mountainous line is currently not clear, particularly when there are photos showing them at the head of the International train in La Paz when there is no chance one of these could have worked such a train up the grade without some substantial assistance. We can only assume they were intended for the more easier grades on the Bolivian Altiplano.

There were no new locomotives built for any of the FCAB owned or operated lines from 1929 until well after the Second World War. During the Second World War however, traffic levels grew (as they did on many lines during this period) and the need for further articulated power resulted in the loan of some Beyer Garratts from Argentina. One of the "Pacific" 4−6−2+2−6−4T Garratts came from the Buenos Aires Midland line to work the Uyuni−Oruro section, and three large 4−8−2+2−8−4T Garratts came from the Cordoba Central Railway.

These latter engines were almost identical to the FCAB engines and were built at the same time, though they were coal fired whereas the FCAB engines were oil fired.

The Aguas Blancas Railway was not converted to metre gauge and continued to operate as a branch on 2ft 6in gauge. Interchange of rolling stock necessitated bogie exchange but this was rarely required as the Antofagasta facilities and Mejillones line and workshops retained the third rail facility. This system struggled on despite the rapid decline in the nitrate trade in the early 1930s. The only engines we have been able to identify as transferring to this line following the 1926−28 gauge conversion are No's 525−9 which were five 2−8−0s rebuilt from the cannibalised remains of Nos 119−128, 2−8−0s originally rebuilt from 2−8−2s from Hawthorn Leslie, and No. 530 from the remains of No's 129−138, the Hunslet batch built to the same drawings in 1907. The branch continued to handle ever decreasing tonnages which saw a drop in its locomotive stock from 8 in 1953 to only 6 four years later. The line closed in June 1961 in which year it was listed as still having 69 miles of track, 3 main line locomotives, 1 shunter, 3 coaches and 332 wagons, all 2ft 6in gauge.

The track was lifted soon after and the buildings in Coloso demolished. The company name however survived into the late 1970s as a 7 mile section of the Aguas Blancas right of way was leased for FCAB track. For accounting purposes the rental received was used to defray local administrative expenses and taxation.

ANTOFAGASTA (CHILI) AND BOLIVIA RAILWAY CO. LTD.

METRE GAUGE LOCOMOTIVE LIST

3. Old Numbering Series for metre gauge locos from 1906 to 1928

This list covers all the metre gauge locos built prior to 1928 as well as some 2ft 6in gauge locos converted to metre prior to 1928, when there was a general renumbering.

3.1 Bolivia Railway Co.

In 1909 the FCAB took over the lease of this railway from Speyer & Co., New York. The Bolivian section locos were lettered FCB but numbered in the main FCAB list. These locos are listed as built for the Bolivian National Railways and were initially numbered accordingly.

Original Numbers	1928 Numbers	Type	Builder	Works Numbers	Year Built	Remarks
1−4	401−404	2−8−0	Alco R	41130−41133	1906	
5−8	405−408	2−8−0	Alco R	44424−44427	1909	

Eight 2−8−0s were built by the Rogers Works of the American Locomotive Co. for the metre gauge Bolivian Railway Company between 1906 and 1909. The first of these—No. 1—is illustrated above. No. 5 (of the second batch) seen below, differs in details with outside valve gear, different safety valve/whistle arrangement and different cab and tender.
Alco Historic Photos (both)

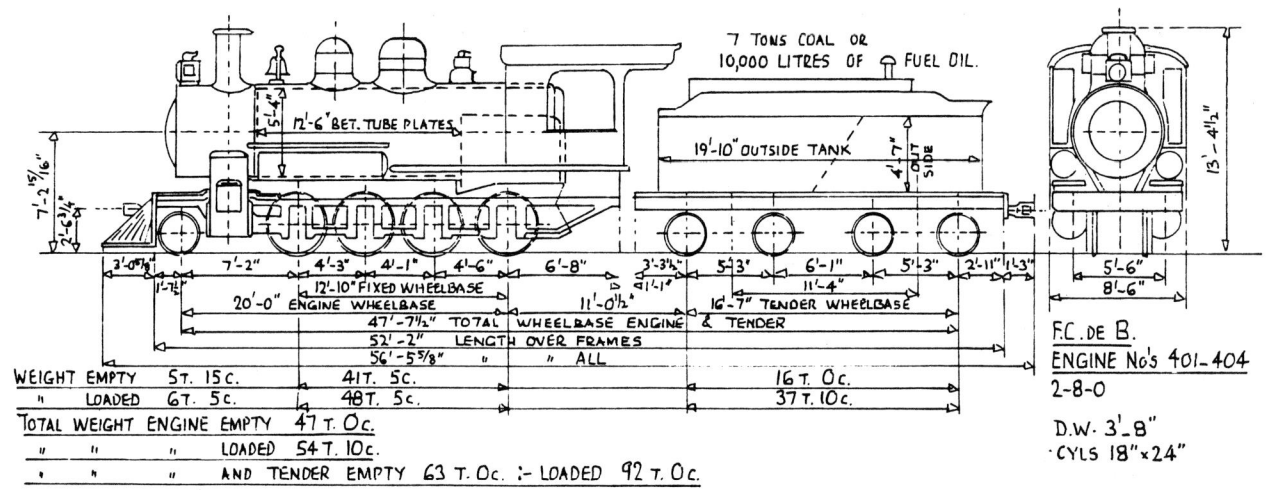

3.2 FCAB Locos built up to 1928 Renumbering

Original Numbers	1928 Numbers	Type	Builder	Works Numbers	Year Built	Remarks
601–602	409–410	2-8-2	K	4860–4861	1912	Rebuilt 1939 with 21in × 24in cylinders.
101–102	411–412	2-8-4T	HE	1102–1103	1912	Supplied lettered FCB.
51–56	451–456	0-6-2+ 0-6-2TT	BP	5617–5622	1913	Meyer type, designed to operate cab-leading. Supplied lettered FCB.
33–36	333–336	4-6-2	Hen	12748–12751	1914	
57–60	357–360	2-8-0	Hen	12544–12547	1914	
61–65	361–365	2-8-0	Hen	18305–18309	1922	
33–52	33–52	2-8-4T	NBL	29562–29581	1927	
84	–	2-6-0	BLW	9859	1889	Converted from 2ft 6in gauge in 1917. Out of service by 1925.
181–182	366–367	2-8-0	Hen	18310–18311	1922	Renumbered 66–67.

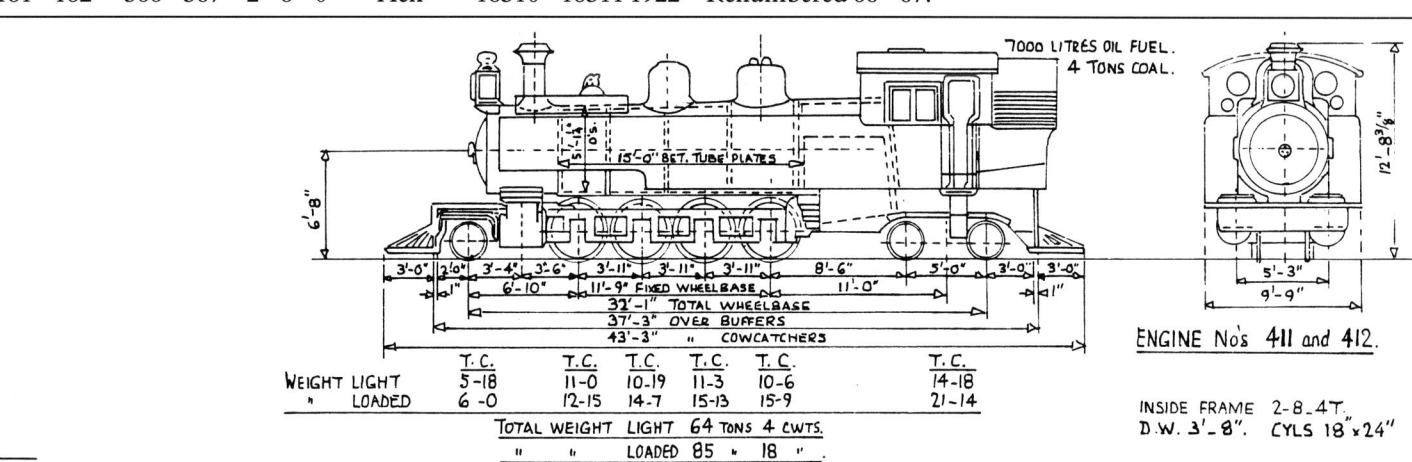

This chunky 2–8–4T – No. 101 – was one of two supplied by Hunslet in 1912 to the metre gauge Bolivian section of the FCAB which account for the FCB initials. Generally conventional in their design, of note are the large oil headlamps front and rear, and the long air reservoir tanks on the top of the side tanks.
Hunslet Engine Co.

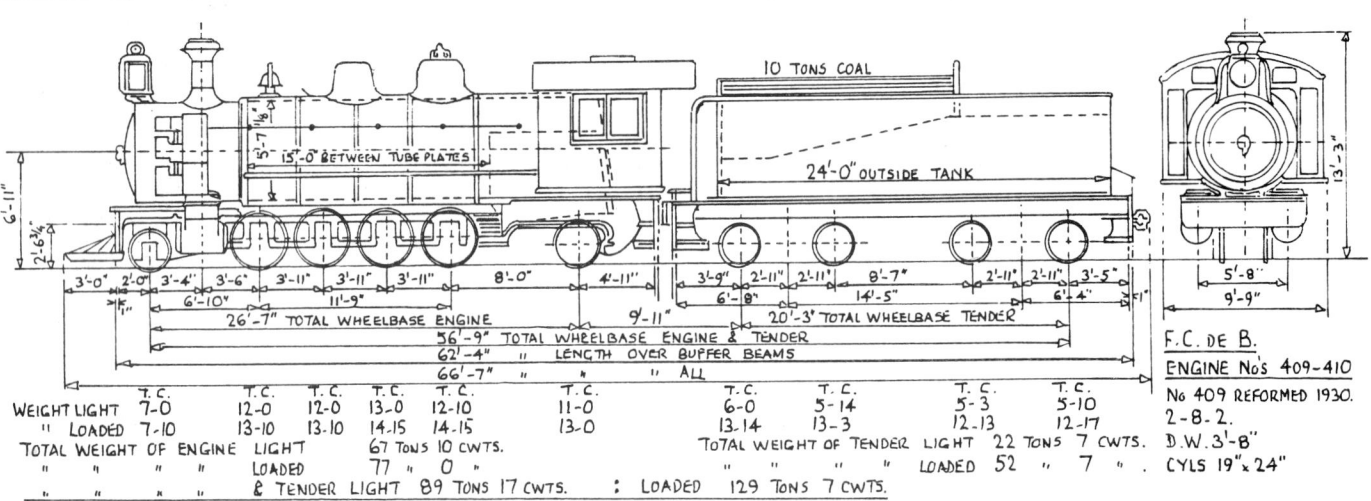

No's 51–56 were designed by the Consulting Engineers Livesey, Son & Henderson and were 0–6–2+0–6–2TT designed to run cab forward. As built they required a crew of at least four, a driver, fireman and two coal carriers who were responsible for carrying the sacks of coal from the separate tender, along the running plate to the cab. Not unexpectedly they were soon converted to oil fuel and proved highly successful.
D. Binns collection

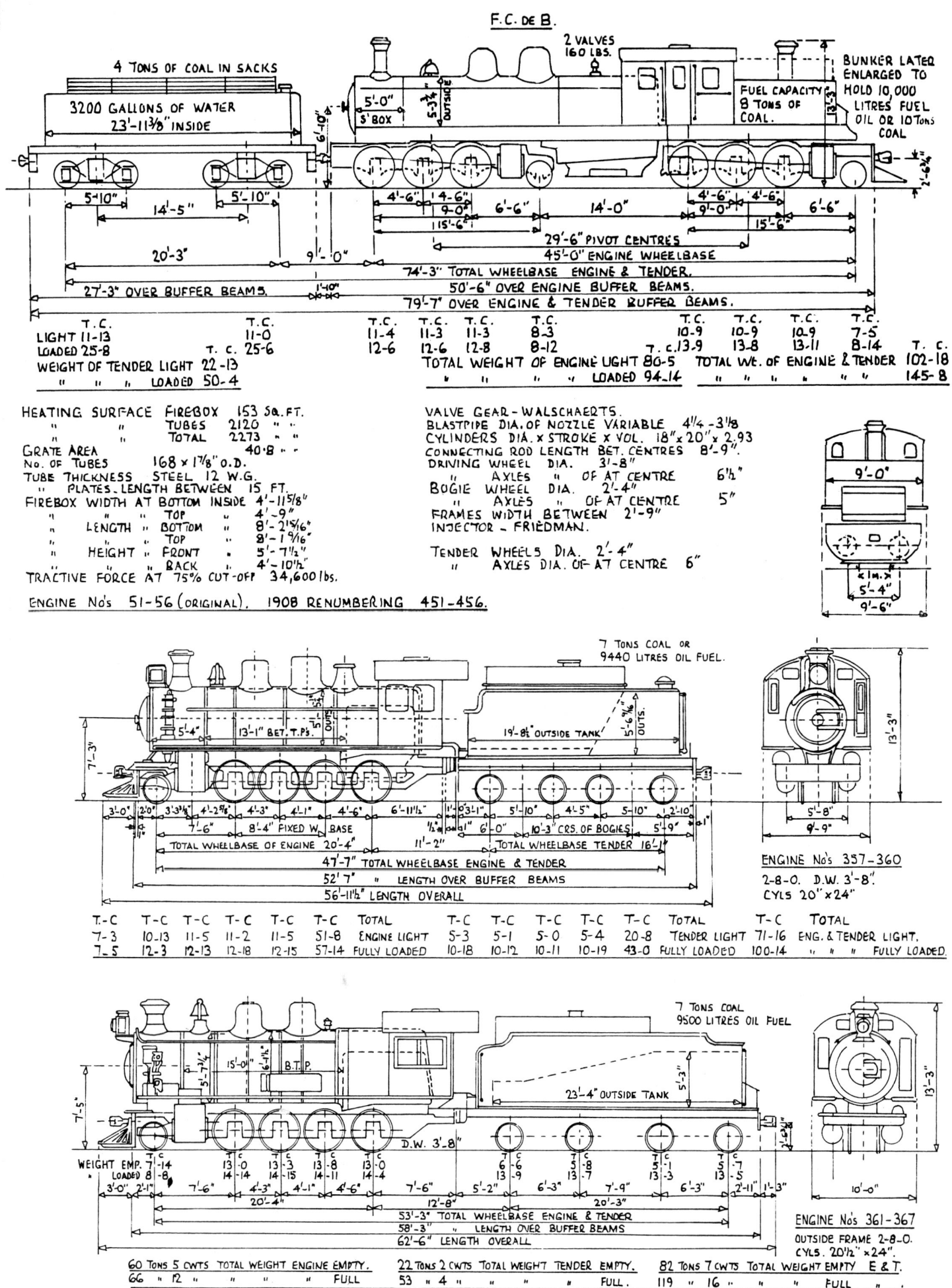

A batch of four 2-8-0s were built by Henschel & Sohn in 1914 taking Road No's 57-60, (later renumbered 357-360). No. 60 is seen in this official broadside view.   U. Bergmann collection

A second delivery of 2-8-0 was made by Henschel & Sohn in 1922 and these locomotives were numbered 61-65 (later renumbered 361-365) and No's 181-182 (later No's 366/7). The class is illustrated by this works photograph of No. 64.   U. Bergmann collection

No. 63 of the 1922 Henschel-built 2-8-0 locomotives is seen here at Atocha on the Uyuni-Villazon (Argentine border) section of the Bolivian State Railways c.1928. This would appear to confirm that FCAB locos did work on this line from time to time. Note the large boiler, feed water heater on the running plate and the Maltese cross on the smokebox door, the significance of which is not known.   Railway Magazine

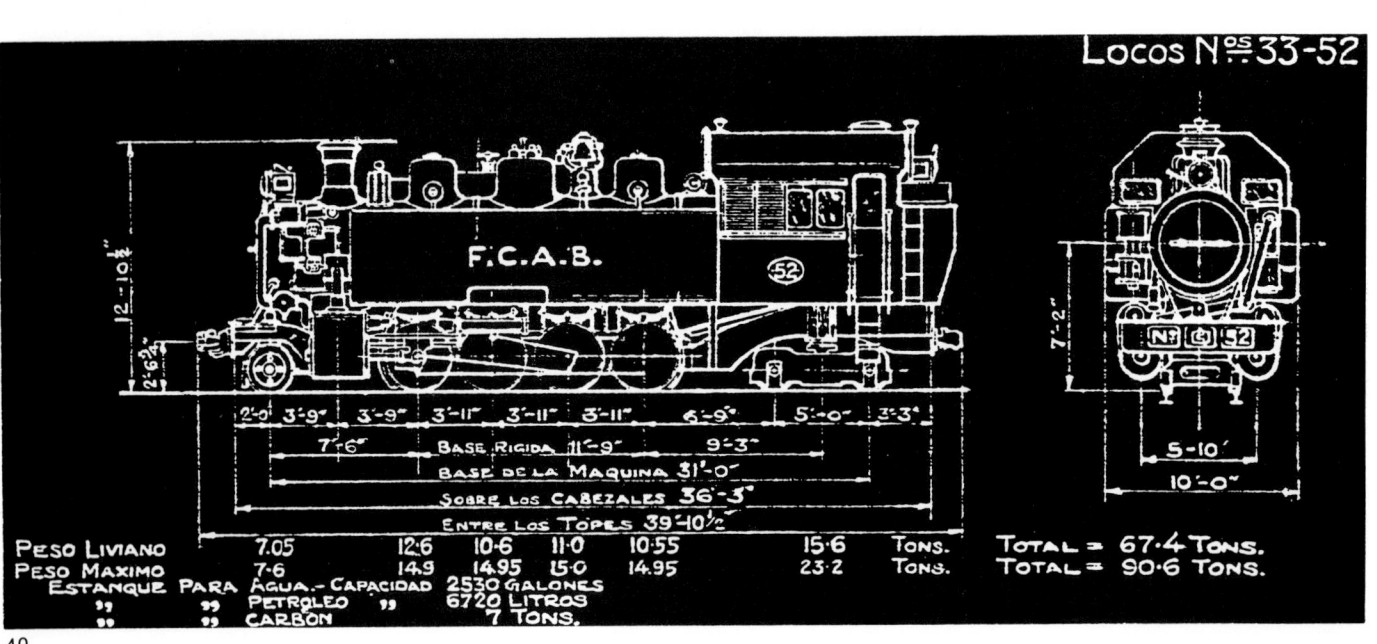

Illustrating the adage "everything but the kitchen sink" is "standard" 2−8−4T No. 38, one of the first batch of these very popular engines that were built by North British in 1927. With 19in×24in cylinders and 44in driving wheels, they weighed just over 90 tons in working order. The oil fuel was carried in a cab-top tank as well as the bunker. Noticeable is the large air brake pump on the front, very necessary on the steep grades of the FCAB. It is recorded that sometimes the oil fuel leaked down the necks of the engine-men!
B Fawcett/J. Buckland collection

Below: No. 43 of the same class.
J. M. Turner collection

In 1914 four 4−6−2 were delivered from Henschel & Sohn, carrying Road No's 33−36, but for what purpose they were intended is not clear. Certainly they could not have worked the heavy International train single-handed, but were probably for use on the more level sections of the F.C. de B. Our photograph of No. 35 illustrates this series.   U. Bergmann collection

3.3 Chilian Northern Railway Co. Ltd.   LOCOMOTIVE LIST

In 1919 the FCAB took over the operation of the CNR Co. They inherited a batch of locos which had previously been the property of the Chilian Longitudinal Railway, some of which were probably used during construction. New locos were ordered for the line by the FCAB and, although numbered in the FCAB 9XX series, were lettered FCNC.

Original Numbers	Second Numbers	Type	Builder	Works Numbers	Year Built	Remarks
1−2	901−902	0−6−4T	HE	1062−1063	1911	These locos were named. Series renumbered 901−904 but whether in original number order is not certain. Some were later fitted with tenders to become 0−6−4TT. 3−4 later became Ferrocarril Estado (Chilian State Railways) No's 3060−3061.
3−4	903−904	0−6−4T	HE	1065/1078	1911	
5−8	1005−1008	2−6−0	Hen	10702−10705	1911	
9−20	1009−1020	2−6−0	Hen	10969−10980	1911	
21−32	1021−1032	2−6−0	OK	5201−5212	1912	
33−38	1033−1038	0−6−0T	OK	5461−5466	1912	
		2−6−0	Hen	10706−10710	1911	Built for the Chilian Longitudinal Railway as part of the batch shown above but taken into FCAB stock.
		2−6−0	Hen	11080−11084	1912	
910−911		2−8−2	NBL	23298−23299	1925	Ordered by the FCAB, numbered in the FCAB series, but lettered FCNC.
912−914		2−8−2	BP	6414−6416	1928	
915−916		2−8−2	YE	2554−2555	1955	

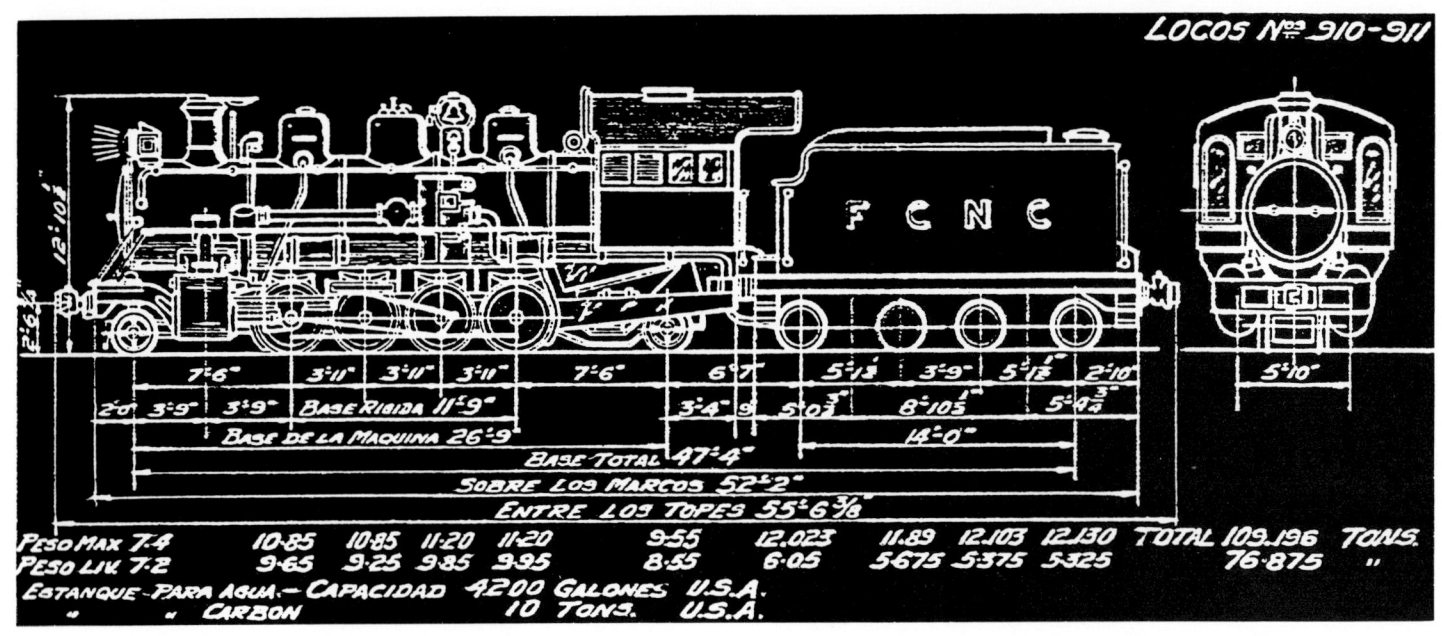

No. 912 of the Chilian Northern Railway was ordered by the FCAB but lettered FCNC. It was one of three built by Beyer Peacock & Co. Ltd in 1928. Note the jacks on the front of the running plate ahead of the cylinders and the large Worthington feed water heater and pump above the driving wheels. D. Binns collection

4. F.C.A.B. Locos from 1928

LOCOMOTIVE LIST

This list includes all metre gauge locos from 1928. Included are those renumbered and converted from 2ft 6in gauge, indicated*. The Bolivian section of the FCAB was nationalised in 1962 to become the Empresa Nacional de Ferrocarriles Bolivia (E.N.FF.CC., now E.N.F.E.) and a number of locos were renumbered into the ENFFCC/ENFE list as shown.

Original Numbers	ENFFCC/ ENFE Numbers	Type	Builder	Works Numbers	Year Built	Remarks
*1		2−8−2T	HL	2950	1912	Converted from 2−8−2 in 1922 (ex No. 168)
*5		0−6−2T	HC	782	1906	Sold to Soc. Fabrica de Cemento, El Melan, 3/1938
*6		0−6−2T	HC	783	1906	Last rebuilt in 1954.
*12−14		0−6−4T	HE	878−880	1905	See footnote to list 2. No. 12 rebuilt as 0−6−0T.
*15		0−6−4T	HE	907	1906	See footnote to list 2.
*20		0−6−4T	HE	912	1906	See footnote to list 2.
*21−26		0−6−2T	HE	945−950	1907	Converted from 0−6−4T. 21 and 25 sold to contractor in Arica c.1962 and still in existence (1990).
*27−28	553−554	2−8−4T	K	4843−4844	1911	
*29−30		2−8−4T	K	4845−4846	1911	Sold to F.C. Taltal
*31−32		2−8−4T	K	4847−4848	1911	

Original Numbers	ENFFCC/ENFE Numbers	Type	Builder	Works Numbers	Year Built	Remarks
33–52		2-8-4T	NBL	29562–29581	1927	
*139–140		2-8-0	HC	787–788	1907	139 scr. in 1960, 140 stored in 1960.
*141–148		2-8-0	HE	958–965	1908	
*150		2-8-0	HE	967	1908	
*151–152		2-8-0	HE	1066–1067	1911	
*153–160		2-8-0	NBL	19428–19435	1911	157 sold to F.C. Taltal
*161–167		2-8-2	HL	2943–2949	1912	
*169–170		2-8-2	HL	2951–2952	1912	
*171–180		2-8-2	Hen	11891–11900	1913	
181–183		4-8-2T	HL	3743–3745	1929	
201–206	821–822 (202/206)	4-8-2	VF	6170–6175	1954	
333	756	4-6-2	Hen	12748	1914	Originally numbered 33.
334–336	751–753	4-6-2	Hen	12749–12751	1914	Originally numbered 34–36
337–338	754–755	4-6-2	Hen	21213–21214	1928	Originally numbered 37–38.
341–346	811–816	4-8-2	VF	6176–6181	1954	
347–348	817–818	4-8-2	VF	6166–6167	1954	
349–350	819–820	4-8-2	VF	6169/6168	1954	
*351		2-4-2	BLW	8215	1886	Cia de Huanchaca No. 20, FCAB Renum. 33 1909–1914.
357–360	614–617	2-8-0	Hen	12544–12547	1914	Originally numbered 57–60
361–367		2-8-0	Hen	18305–18311	1921	Originally numbered 61–65 and 66–67 (Ex 181–182).
390	909	4-8-2+2-8-4T	BP	6524	1928	Beyer Garratt. Originally numbered G1
391–392	901–902	4-8-2+2-8-4T	BP	6525–6526	1928	Beyer Garratts. Originally numbered G2–G3.
393–398	903–908	4-8-2+2-8-4T	BP	7420–7425	1950	Beyer Garratts.
401–404	606–609	2-8-0	Alco R	41130–41133	1906	
405–408	610–613	2-8-0	Alco R	44424–44427	1909	
409–410		2-8-2	K	4860–4861	1912	Rebuilt 1939 with 21in × 24in cylinders.
411–412	551–552	2-8-4T	HE	1102–1103	1912	
451–456		0-6-2+0-6-2TT	BP	5617–5622	1913	Meyer type. Originally numbered 51–56.

This attractive 2ft 6in gauge 2-8-2, No. 169, was one of a batch of ten delivered by Hawthorn Leslie in 1913. These engines were standardised with a 2-8-4T type delivered the year before by Kitson. They had 17in×22in cylinders, 37½in driving wheels, and weighed 87 tons in working order. One of the class was later converted to a 2-8-2T, and all survived to be regauged to metre and led long useful lives.   *J. M. Turner collection*

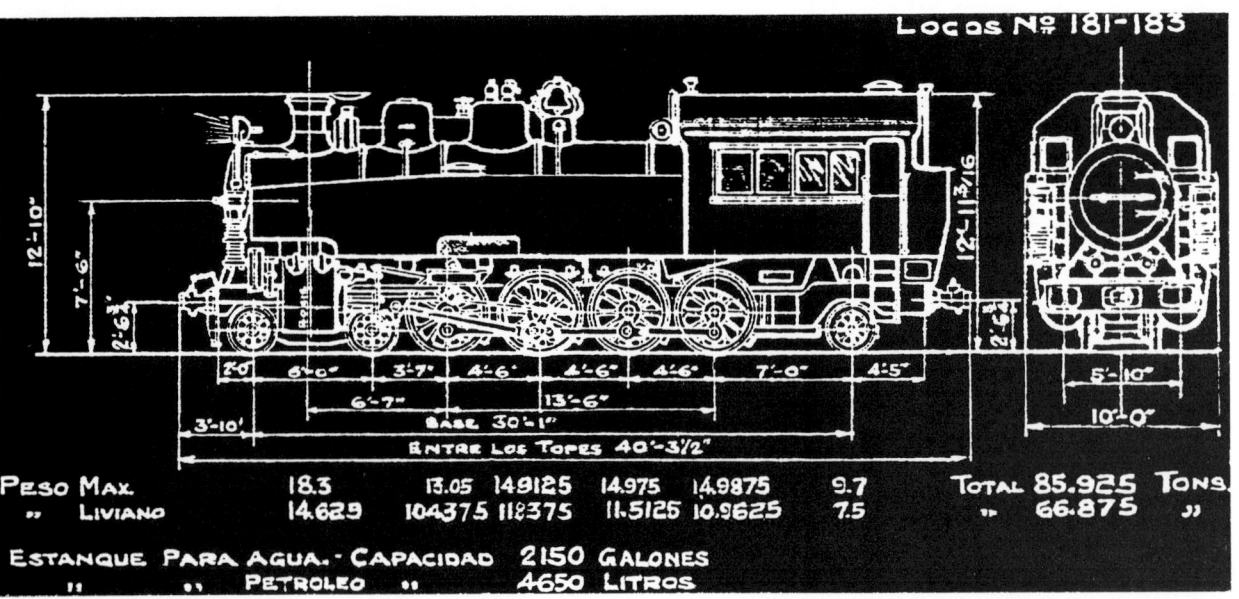

Highly polished 4–8–2T No. 182, built by Hawthorn Leslie in 1929 for passenger duties, stands at the head of the six car International Train at the new Antofagasta station in March 1939. The apparently small vehicle at the end of the train is a 2ft 6in gauge coach converted to metre gauge and appears to be a business or inspection car. *B. Fawcett/J. Buckland collection*

No. 182 was one of a series of three 4–8–2Ts built by Hawthorn Leslie in 1929 and classified as "passenger engines". They were similar to the earlier standard North British 2–8–4Ts and shared the same boiler. Their business-like appearance included 21in×24in cylinders and 51in driving wheels. The oil fuel was carried in a bunker above the cab. Note the highly polished appearance of the loco, a feature of most FCAB engines in Company days. *B. Fawcett/J. Buckland collection*

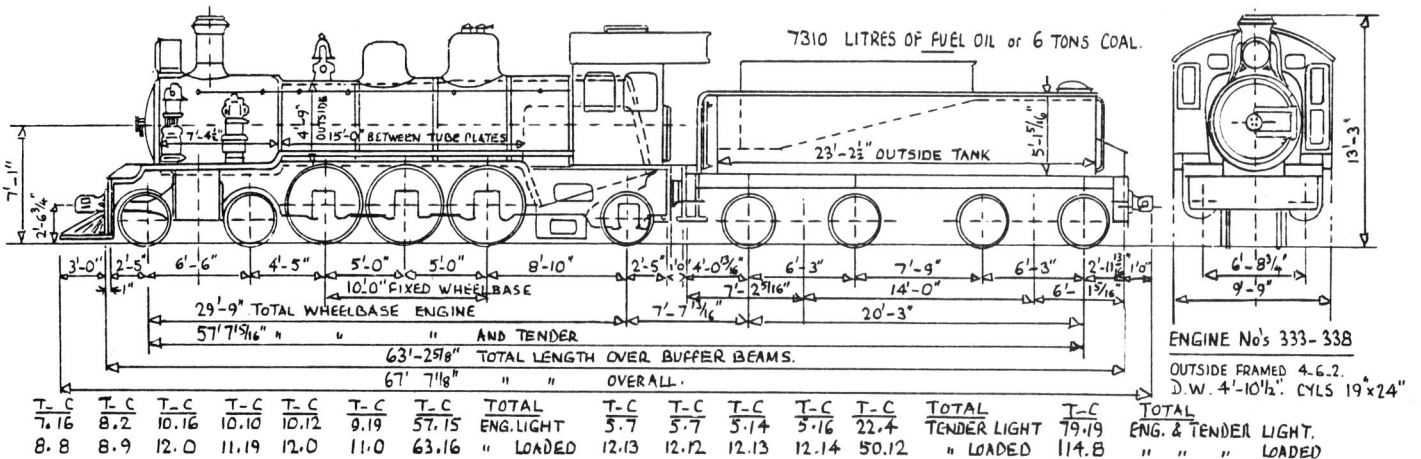

No. 337 was the first of a pair of 1928 Henschel-built 4—6—2 originally numbered 37 and 38. They were eventually to become No's 754/5.   U. Bergmann collection

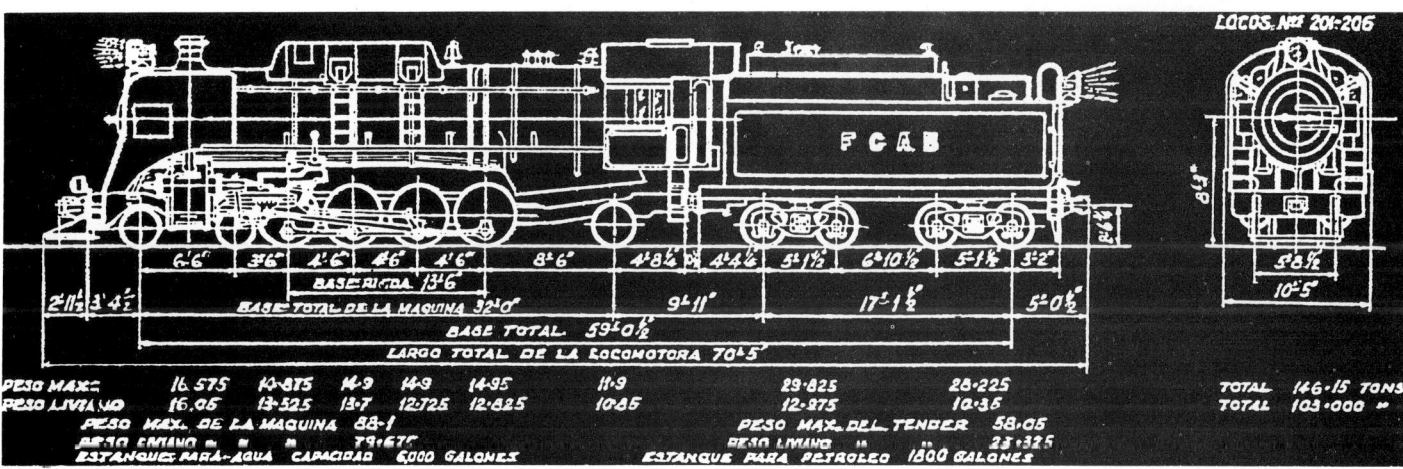

Opposite page—bottom photograph: This interesting view, taken on 13 September 1929, shows Beyer Garratt 4—8—2+2—8—4T No. 391 with a long passenger train at Chijini station, La Paz. The engine is probably on its trial run and was the first of three Garratts delivered from Beyer Peacock in 1929. Six more followed in 1950.   R. F. Ellis collection

Diagram of the 1950 design of Beyer Garratt for the FCAB, six of which were built in Manchester. These carried Road No's 393—398 (later 903—908).

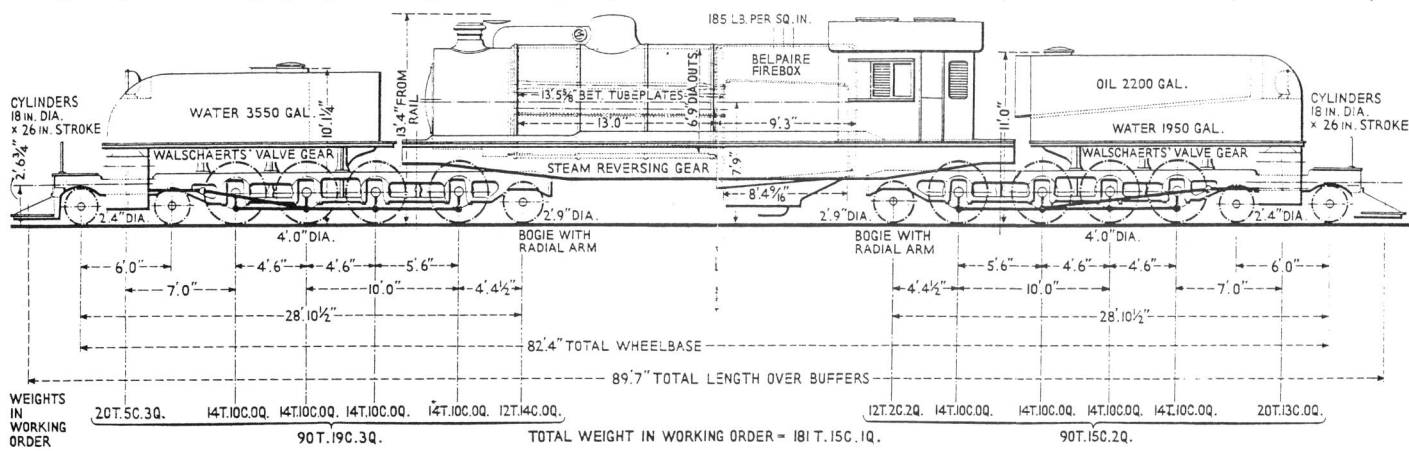

During the Second World War two Argentinian metre gauge railways loaned Beyer Garratts to the FCAB:
Buenos Aires Midland Railway:- 1 loco from:

101–102	4-6-2+2-6-4T	BP	6570–6571	1929	Worked on the Uyuni–Oruro section.

Cordoba Central Railway:- 3 locos from:

1511–1520	4-8-2+2-8-4T	BP	6550–6559	1929	These were almost identical to the FCAB Garratts, but were coal fired whereas the FCAB locos were oil fired.

The following are shown in the Henschel works list as supplied to the FCAB. Whether they were actually for the FCAB, one of its subsidiaries, or a mining company, is not known:

0-6-2T	Hen	21014	1928
0-6-2T	Hen	21280	1929

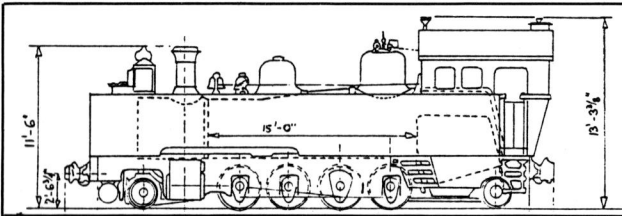

THE FCAB CIRCA 1930

A detailed look at the FCAB line is presented in the form of excerpts from a travellers booklet published by the Company in 1930.

"The Antofagasta Railway connects the Chilian ports of Antofagasta, Mejillones and Coloso with the principal towns in Western Bolivia and with the Argentine and Peru. It affords access to some of the finest mountain scenery in South America, bringing the romantic land of the Incas within easy reach of the tourist by means of its International trains, provided with sleeping carriages and restaurant cars equipped with every luxury.

The main line starts at Antofagasta, a town with some 70,000 inhabitants on the Chilian Coast, and 590 miles north of Valparaiso. It is reached from England in about 26 days via the Panama Canal, and in about the same time via Buenos Aires, and thence by rail to Bolivia and over the Company's own system from Atocha. Another route which takes a few days longer is via Buenos Aires and the Chilian Transandine Railway. The traveller then has the choice of continuing his journey to Valparaiso and making a short sea trip to Antofagasta, or the latter may be reached by rail via the Chilian Longitudinal Railway to Baquedano and thence over the Company's line.

Through sleeping trains de luxe convey passengers from Antofagasta to La Paz, and the trains are timed to reach La Paz during the afternoon, in good time to appreciate the views of the magnificent snow capped "Illimani" and the unequalled panorama which unfolds itself of the city of La Paz when the train leaves the flat tableland at the "Alto" station and proceeds down the incline to the charming old Spanish type of city.

From Antofagasta the railway (having to reach an altitude of 13,000 feet in 227 miles) loses no time in beginning its climb, and at Portezuelo, 29 kilometres (18 miles), the rail level is already 1,800 feet above the sea, giving an average grade of 1in50 but in several places it is as steep as 1in30. At O'Higgins, kilometre 36, is the junction of the branch to the "El Boquete" nitrate fields. The end of this branch is some 5,640 feet above the sea. At Baquedano, kilometre 96, the Chilian Northern Longitudinal Railway crosses the main line of the Antofagasta Company. . . . At kilometre 117 the main line enters the principal nitrate district of this part of Chile and leaves it at kilometre 172. In this section are situated some 24 oficinas (nitrate factories). . . . After leaving the Nitrate Zone we catch out first view of the Andes, and, soon after, cross the river Loa and reach the picturesque little town of Calama at kilometre 239 (149 miles) from Antofagasta.

Here the eye is refreshed with green pasture lands, irrigated by the waters of this river, for till now, the line has passed through what is to all appearances a barren district without a blade of grass to be seen anywhere, though the hills on either side are not without a peculiar beauty of their own, due to the variegated colouring caused by the presence of rich copper ores. The International train stops at Calama for about 20 minutes, of which the traveller will eagerly avail himself for a stroll in the brisk air, should it be in the warm season. In the Winter, he will perhaps prefer to remain in his comfortable, specially-warmed compartment, and view from there the magnificent snow-capped Andes.

Calama is a town some 7,450 feet above the sea, and some passengers to Bolivia prefer to stop for at least a day here to accustom themselves to the altitude before going further, but beyond slight headaches and a shortness of breath, passengers in normal health have nothing to fear from the altitude.

At kilometre 254 is the short branch (10 kilometres long) up to the copper mines at Chuquicamata, 8,846 feet above the sea. These mines, which in reality comprise a prolonged succession of hills, have been acquired in great part by the Anaconda Copper Company. A special process for dealing with low-grade copper ores by means of electrolysis is employed, and on a scale hitherto unknown. The production of pure copper, when all the plant is working, is estimated at 600 tons a day, from ore of about 1½ per cent ley, so that Chuquicamata has become one of the chief sources of the world's supply.

A visit to this vast and unique establishment, with its railways, its steam excavators, its depositing tanks, and appliances capable of handling some 40,000 tons of copper ore a day, will not fail to be of the greatest interest to the traveller.

The next point of interest is met near Conchi station, some

300 kilometres from Antofagasta, where a deviation, 12 kilometres long, has been made to avoid the two wrought-iron viaducts which previously carried the railway over one dry river bed and also the river Loa. The larger viaduct, which is a most graceful structure, has six lattice girder spans of 80 feet each, supported by Phœnix column towers, and a fine view of this structure is obtained on leaving Conchi station. The line drops down at an easy gradient, passing through deep cuttings in extraordinary volcanic deposits, and crosses the river Loa on a high embankment, pierced by a stone culvert through which the river flows.

From Conchi station runs the branch line (20 kilometres long) to the copper mines of Conchi Viejo, the rail level at the end of this branch being 11,453 feet above the sea. At San Pedro station, kilometre 318 (197 miles) and 10,630 feet above the sea, are situated the collecting reservoirs, blasted out of solid rock, of the Waterworks which the Antofagasta Railway Company has constructed. The Company has expended approximately £1,930,000 in order to supply the town of Antofagasta, the nitrate fields and its own services with water, for no other fresh water can be obtained except by condensing sea water.

From these reservoirs pipes are laid the whole distance of 197 miles, delivering the pure snow water of the Andes at the sea level, which is no inconsiderable undertaking in itself. The water to fill the reservoirs is taken from three sources, including the Siloli spring, some 60 kilometres to the northeast of the railway line, at an altitude of some 14,500 feet above the sea. This source provides 7,500 tons of water a day, through pipes of 11 inches diameter. The chief source of water supply is thus conducted to Antofagasta through pipe lines some 235 miles in length.

Shortly after leaving San Pedro station, the Railway skirts the bases of the majestic, snow-capped volcanoes "San Pedro" and "San Pablo". From the crater of the former ascends a constant column of smoke, and, though it has not shown greater signs of activity than this in recent years, it is evident that in comparatively modern times it has been in eruption, for the railway cuts through a lava bed nearly a third of a mile wide, which looks as fresh as if it had been deposited only a year ago.

Climbing steadily up, at Ascotan; kilometre 366 (227 miles from Antofagasta), the summit of the Main Line (Chilian Section) is reached at a level of 13,000 feet above the sea, and from here it descends rapidly to a level of 12,256 feet at Cebollar, kilometre 393, where it runs alongside a wonderful lake of borax, some 24 miles long, owned and until recently worked by the Borax Consolidated Company.

The view as the train winds round the snow-capped mountains, whose slopes are bright with metallic hues, is quite unique, and the glistening surface of the borax lake, with occasional stretches of verdant water, remind one forcibly of Switzerland. From Cebollar station a short branch runs into the calcinating establishment. This lake is said to be the largest single deposit of borax in the world.

At Ollagüe station, kilometre 442, is the junction of the branch line (95 kilometres long) which was constructed in 1907 to serve the important group of copper mines at Collahuasi, said to be among the richest known.

This branch and the Potosí line are believed to be higher than any other lines of railway in the world, for their rails reach to

the great heights of 15,809 and 15,705 feet, respectively, above sea-level, and by those whose respiratory organs do not suffer from the altitude, the Collahuasi branch is well worth a visit, not only on this account, but also because of the truly magnificent panorama of snow-clad mountains to be seen on the journey, among them the giant "Ollagüe", upwards of 20,000 feet high. This is the one part of the Antofagasta Railway where snowstorms are troublesome.

Shortly after leaving Ollagüe station, the frontier line between Bolivia and Chili is crossed at kilometre 444 (276 miles from Antofagasta), and from this point to Uyuni

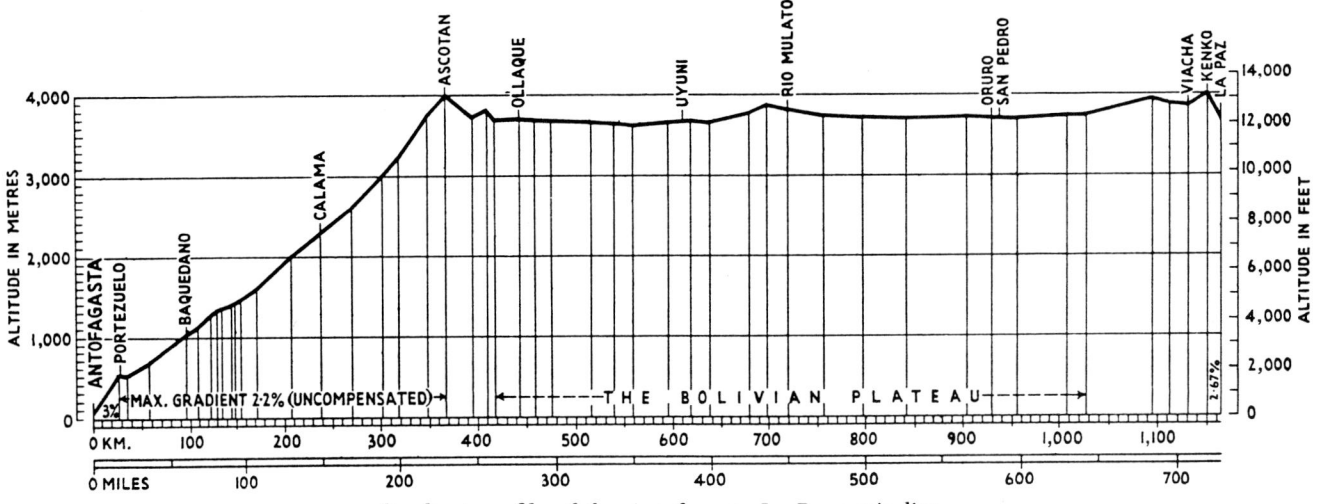
Gradient profile of the Antofagasta-La Paz main line

(kilometre 617) the line runs at an almost uniform level of 12,000 feet above the sea over the Bolivian tableland, which stretches for hundreds of miles on each side of the railway. The plain around here is sandy and full of salt.

Uyuni is a town of 5,000 inhabitants, many of whom are Indians. Here, for the first time, we make the acquaintance of the llama as a beast of burden. Until the opening of the Potosí and Atocha branches these animals were largely employed in the transport of tin and silver ores. They used to travel from Potosí in troops of 100 or more and took 15 days to complete the journey of some 125 miles. The load for each animal is 100lbs, and it is said that if weight is exceeded, however small, the llama will refuse to stand up. Uyuni is of comparatively recent growth, having been founded principally because at one time it formed the railway terminus, and some of the workmen consequently had to live there.

The private railway, 33 kilometres long, of the Huanchaca Company runs from here to Pulacayo, a mining town of about 8,000 people, situated at an altitude of 13,600 feet. The mines belong to a Franco-Chilian company, and those who visit them, find them extremely interesting.

The Bolivia Railway Company's branch, of metre gauge, runs from Uyuni to Atocha, a distance of 90 kilometres (56 miles), and serves for transporting the output of the rich mines at Quechisla. It also forms part of the connecting link with Buenos Aires, and many travellers now use this route in preference to the older one via Antofagasta, Valparaiso, Los Andes and Mendoza. The journey from Antofagasta to Buenos Aires occupies four days, dining and sleeping coaches being provided throughout the journey of 2,700 kilometres. The service is open all the year round, and there are no stoppages on account of snow.

Trains for Potosí run direct from Uyuni, the junction with the Potosí line (174 kilometres in length) being at Rio Mulato, kilometre 722 of the main line. To reach the historic city, the train climbs to a height of 15,705 feet at Condor, or, say three miles above sea level. All this part of Bolivia is very broken and wind-swept, and often covered with snow, and one cannot fail to be impressed with the difficulties which had to be surmounted when the line was built. The track winds through deep cuttings in solid rock and along the side of mountain after mountain until Potosí is reached.

At Agua de Castilla are the ancient Porco mines, which yielded silver before the famous Potosí deposits were discovered.

Potosí was founded in 1546 by Don Juan de Villaroel Santandía, and his title of discoverer and founder was confirmed in the following year by the Emperor, Charles V. The mountain itself, which is 15,900 feet high, is famous, and legends regarding its discovery are many. One of them, probably the most authentic, is to the effect that an Indian was sent from Porco to look for some llamas which had strayed, and that he found them on the conical mountain now called Potosí. As it was too late for him to return to Porco on that evening, he tethered the llamas to some scrub, and in the morning found that a piece of the scrub had been torn up by the roots, to which little silver threads were hanging. On his return to Porco he informed Villaroel of his discovery.

. . .The first mint in South America was erected here, . . . the machinery, even to the screws, being all of wood from the Argentine province of Tucumán, and having been transported to Potosí by Indians. . . .

Not far from Potosí is Sucre, the old-time capital of the country, and a Government railway about 175 kilometres long, to connect the two cities is being constructed.

To continue our journey to La Paz, it is necessary to return to Río Mulato.

At Huari, kilometre 807, we come in sight of the mysterious fresh water lake of Poopó, which receives 212,000 cubic feet of water per minute, whereas the outflow during the same time is only 2,000 cubic feet.

From Huari onwards the stations are situated mostly at mining centres, and at Machacamarca, kilometre 906, the private railway, some 100 kilometres in length, built by Señor Simón I Patiño, leaves the Antofagasta Company's main line to connect with the Uncía and Huanuni mines, two of the great tin mines of Bolivia.

At kilometre 930, or 578 miles from Antofagasta, is Oruro, where there is a stop of 20 minutes.

Oruro is a mining town of 28,000 inhabitants, and is 12,160 feet above the sea. It possesses two pleasant "plazas", and has straight streets, where modern buildings stand side-by-side with those of old Spanish design. There is an eighteen-hole golf course, situated about a mile from the station.

Oruro was founded in 1595 by a Spanish priest who discovered valuable silver deposits in the neighbouring hills.

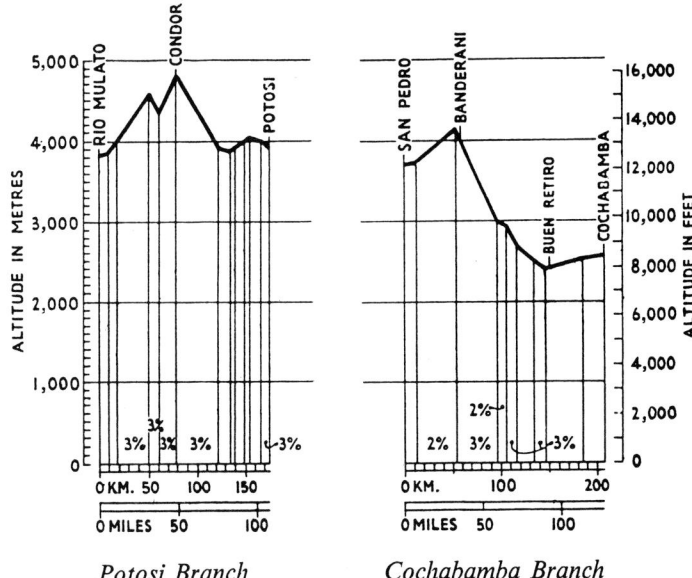

Potosi Branch   *Cochabamba Branch*

As it increased in extent and importance it came, in the time of Phillip III, to be oficially named "Real Villa de San Felipe de Austria", and was second in importance only to Potosí. In September, 1826, the name was changed to Oruro.

Travellers wishing to go to Cochabamba change at Oruro.

At San Pedro, 6 kilometres from Oruro, on the railway to La Paz, the Cochabamba line (128 miles long) branches off, and begins its ascent of the mountains, which have to be crossed before the Cochabamba Valley can be reached. This was another very difficult line to construct, and at one part of the descent, it can be seen in five different places as it winds backwards and forwards on its downward way. The highest point, "Cuesta Colorada", is 13,573 feet above the sea. As the line descends the air becomes milder and milder, until at Arque, kilometre 124, we find lemon and other fruit trees, at an altitude of 8,858 feet.

Cochabamba was founded, under the name of Oropeza, on the 1st January, 1574, and its inhabitants figured prominently in the struggle for independence in the early years of the 19th century. The climate is agreeable, and beneficial to those who have been living for any length of time in the colder and loftier parts of the country. . . .

Outlying suburbs are reached by means of motor-cars and electric trams.

The first section, some 250 kilometres in length, of the railway from Cochabamba to Santa Cruz is now in course of construction. This line is being built by an American Corporation for account of the Bolivian Government, and will eventually give rail connection with Eastern Bolivia, where none exists at present.

From Oruro to La Paz, the first part of the journey is over a partially barren plain, which gradually becomes fertile, and flocks of sheep and llamas, and sometimes a few vicuña, are to be seen grazing leisurely upon the short scrub. Huts, some circular and some oblong, indicate the dwelling places of the cultivators of this region. The Indians continue to till the land in the primitive manner and under the communal conditions of their forefathers.

Viacha, 1,132 kilometres from Antofagasta, is a junction for the Mollendo—La Paz line and that from Arica to La Paz.

After leaving Viacha the line continues to run over flat and stony ground as far as the "Alto", where an elevation of 13,134 feet above sea level is reached, this being the highest point on the main line between Antofagasta and La Paz. From here the line begins to descend to the city of La Paz, 17 kilometres distant. La Paz is the seat of the Bolivian Government, and the railway station is 12,143 feet above sea level. Like the railways to Cochabamba and Potosí, this part of the journey exhibits yet another example of engineering skill and daring.

On reaching the "Alto", glimpses are obtained of La Paz and the suburbs of Calacoto and Obrajes, while immediately after leaving the station the whole of the city and surroundings, nestling in the valley more than 1,000 feet below, bursts suddenly on the view—one of the most wonderful sights in the world.

Everything in La Paz is picturesque. Its situation; its old Spanish dwellings, rambling in varied outline over the hilly surface of the town's site; its troops of llamas and herds of donkeys driven by Indians in gaily-coloured "ponchos"; its ancient churches with beautifully carved portals; its markets, where "cholas" squat on the ground surrounded by their wares; whilst in the distance the "Illimani" rears its lofty and snow-covered peak, as if to constitute itself the guardian of it all.

"Illimani" is over 21,000 feet high, and its name means "the place where the sun rises".

From La Paz excursions can be made to the prehistoric ruins of Tiahuanaco. These ruins consist of immense blocks of stone (one of them measures 29 feet by 15 feet by 5 feet), and one wonders how they were brought to their present sites, because the nearest spot where stone is available is 30 or 40 miles distant.

Not far from Tiahuanaco is Lake Titicaca, which is 138 miles long and 60 miles broad at its widest part, and is the largest lake in South America. Steamers convey tourists from Guaqui, a port on the lake, to different points of interest, such as Copacabana, where there is a renowned shrine, and the islands of the Sun and of the Moon, regarding which there are many legends in connection with the Indians.

The rush boats of the Indian fishermen, which may be seen on the lake, are supposed to be modelled after those used centuries before the Incas founded Cuzco, the capital of their wonderful Empire."

Railway Magazine for June 1928 published an article by Charles Travis, M.Inst.T. entitled "By Railway to the Roof of the World" from which the following paragraphs are extracted, courtesy *Railway Magazine*;

"In the course of an extended tour which I recently made over the railways of South America, I reached Antofagasta after a long train journey of 1,072 miles, occupying close on three days, on the Chilean Longitudinal Railway from Santiago. The connection for La Paz leaves the same night, and unless one travels on that, an enforced stop of four days at Antofagasta is necessary. I stopped over and had a most interesting time on the 2ft 6in gauge lines in and around the port. I had expected to find something in the nature of a toy railway, but was soon to discover that, on the contrary, it was extremely well-equipped system, capable within reasonable limitations of all the work performed on the broad-gauge lines.

The Antofagasta yards have a capacity of 1,300 wagons. . . .

At the end of one of the yards I saw a long freight train

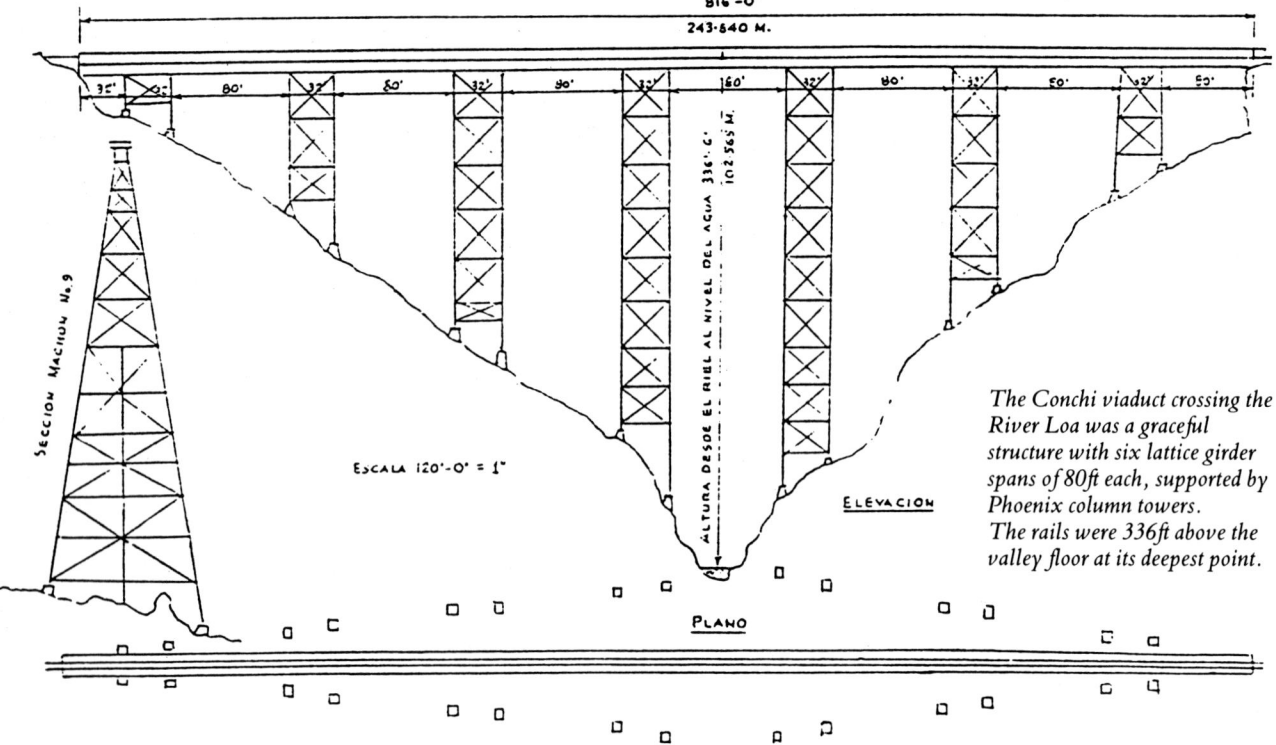

The Conchi viaduct crossing the River Loa was a graceful structure with six lattice girder spans of 80ft each, supported by Phoenix column towers. The rails were 336ft above the valley floor at its deepest point.

steaming out, and was surprised, on counting up the vehicles, to find that the gross load represented 1,500 short tons—and this on the 2ft 6in gauge! The freight vehicles are excellent, the standard being a 20 ton wagon, with new 28 ton wagons being placed in service. In connection with goods traffic the railway administration does not perform the work of collection and delivery, the traders having to bring the goods to the cars for loading, and, in the reverse direction, to come to the station for their traffic. The railways do not even advise the traders of the arrival of traffic for them, this being left as a matter between the consignor and consignee. . . .

I travelled to Mejillones and back on a Drewry rail car driven by a native who didn't believe in losing time. The outward journey was quite fast enough to test the nerves of a seasonal traveller, but it was nothing compared to the mad lick back in the dark. I had been warned of a probable delay at Pampa, where the ordinary train crosses and surely enough there we were delayed. We arrived there at 5.35pm and it was not until 6.50 that we were able to resume our journey. . . .

About 6.15, however, over in the far distance I saw the train approaching, the Pyle headlight flashing clearly over the barren desert, and in due course the mixed train arrived. There were only two cars on it, the rest being wagons of grain, &c. And so, having got the staff we resumed our dash down to the sea. I expected every moment to be my last, as the driver took all the nasty curves at a speed that was really frightening. But we kept the road, and I was able, though bruised and shaken, to keep an appointment to dine with the Resident Engineer of the railway and some of his colleagues. . . .

A section of the timetable showing the international train service between Antofagasta and La Paz is reproduced, and from this it will be seen that the journey of 725 miles occupies 42 hours, but it is made in every comfort, as the trains are provided with excellent dining and sleeping cars.

The train leaves Antofagasta in the evening, and so until daylight dawns one can see nothing of the country, and, as a matter of fact, there is then very little to see. After leaving the coast the line crosses the extensive nitrate district in the Province of Antofagasta, thence through the copper mining centre of Calama. Here and there we passed railway settlements, but that apart there was simply nothing but sand, and in the far distance mountains and high hills. It was desolate in the extreme. At Ascotan, 223 miles from the coast and nearly 13,000ft high, I saw llamas for the first time and a little further on, passed many skeletons and dried corpses of donkeys, horses and other animals. They were lying all over the place, quite close to the railway, yet even where they were near settlements nobody apparently bothered to dispose of them. Surely, it could not be healthy, but there it was. The line later skirts the well-known borax deposit at Cebollar, and thence continues across the Great Bolivian tableland. . . .

We arrived at Uyuni towards 8pm, and there it was necessary to change from the 2ft 6in gauge to the metre gauge. This train was composed of excellent stock with dormitorio (sleeping car) and comedor (dining car) in the formation, and there was no lack of room in the berths. Oruro was reached just before 7 o'clock the next morning, and there another change had to be made—this time to an ordinary compartment car of small merit. There were not even arm rests in the first-class coaches to relieve the strain of a seven hour journey. But there was a comfortable dining car on the train, and I managed to retain a seat there. . . .

I shall never forget Oruro—the gaudily-attired Indian women, all squatting on the ground with the most mixed sort of food you ever saw, the Indian males handling all kinds of bulky packages on their humps, the soldiers all looking very important, and, of course, fully equipped with swords and revolvers; and, altogether the general ensemble.

The diagram reproduced on this page shows the line of deviation 11.78km in length, made in 1918 to avoid two wrought iron viaducts which carried the railway over a dry river bed and across the River Loa.

On page 75 it is stated that "In 1918 a 4.8km deviation was built to avoid the Conchi viaduct" (from FCAB official sources), whilst above, Juan Araya quotes "11.78km". The distance variation is probably explained by rebuilding and relocations to the line since 1918. (Mr. Araya's material is contemporary).

J. ARAYA C.

The route crossing the 336ft high Loa Viaduct was closed when the new deviation was completed. This photograph is dated 24 January 1918 and shows the letting of water through the northern face of the 17 metres high Loa culvert during construction.
D. Binns collection

Oruro is a widely-flung town, apparently of considerable importance—presumably from the mining point of view—and it was some time before we again reached the barren desert country that appears to be the general run out there. We were travelling at a height of 12,500ft above sea level just beyond the township, and I was relieved, some time later, to see a trace of grass on the ground. Barren sand had hitherto been the whole of the vista, and it was a great relief to see something approaching verdure, although it was coarse.

The train from Oruro to La Paz comprised a flat car, conveying a motor car, two cattle wagons conveying horses, a postal van, three second-class cars, two first-class cars, two sleeping cars, one dining car and a brake van—and it ran jolly well!

As the journey proceeded we reached quite good country, and saw a lot of sheep and cattle, while there was on both sides of the line, a thin verdure of grass. Very soon afterwards, we came across several semi-fortified places, in each of which were huts, apparently tenanted by the Indians. They looked more like mushrooms, with their thatched tops, and their sloping sides. What the outer walls were for I could not imagine, 'cos there were so many holes in them. They were weird. Surrounded by scores of dogs—there's no dog tax in that Continent—the living places are of the most peculiar construction—and they apparently house multitudes of people.

Far to the right, as we travelled on, we could see a great, gaunt mountain, which must have been a tremendous height. I found later that it was Illimani, the second highest peak in South America. Cone-shaped, it was entirely covered with snow, and, as I found out later in La Paz, it dominates the city. A wonderful sight! Getting nearer La Paz we entered rocky country, such as I have seen nowhere else in the world. Towering cliffs, scoured out into all conceivable shapes and sizes, presumably by the effluxion of time and weather, stood out in bold outline against the clear blue sky. There below, was the city, with its red tiled roofs, looking more like doll-houses....

It seemed almost impossible for a train ever to negotiate the precipitous slopes, but by a wonderful series of loops and curves, we eventually worked our way down from Viacha, facetiously termed the "Crewe of Bolivia", through Kenko, over the edge of the world, and down to the old-world city of La Paz, with its two stations of no architectural merit.

Viacha station circa 1925.   *D Binns collection*

Antofagasta (Chili) and Bolivia Railway Co. Ltd.

18726 Imp. Macfarlane, Antof

INTERNATIONAL TRAINS BETWEEN CHILI & BOLIVIA

(SUBJECT TO ALTERATIONS)

Direct Service between Antofagasta (Chili) and La Paz (Bolivia) in 42 Hours.

Altitude above sea level (metres)		STATIONS		ARRIVE	DEPART		STATIONS		ARRIVE	DEPART
		Antofagasta	Tuesday & Saturday		19.50		La Paz	Tuesday & Friday		15.30
2272		Calama	Wednesday & Sunday	5.02 C.T.	5.30 B.T.		Viacha	,, ,,		17.02
3702		Ollague	,, ,,	14.03	15.03	Cochabamba branch	Cochabamba	Tuesday & Friday		7.30
3660		Uyuni	,, ,,	20.00	20.30		Oruro	,,	19 30	
	Tupiza branch	Uyuni	Saturday & Wednesday		8.40		Oruro	Tuesday & Friday	22.21	22.40
3648		Atocha	,, ,,	12.09		Potosi branch	Potosi	Friday		7.00
3806		Rio Mulato	Thursday & Monday	0.09	0.24		R. Mulato	,,	16.40	
	Potosi branch	R. Mulato	Thursday		10.18		Rio Mulato	Wednesday & Saturday	4.42	4.54
3905		Potosi	,,	19.34		Tupiza branch	Atocha	Wednesday & Sunday		14.30
3696		Oruro	Thursday & Monday	6.40	7.30		Uyuni	,, ,,	18.40	
	Cochabamba branch	Oruro	Thursday & Monday		8 30		Uyuni	Wednesday & Saturday	7.48 B.T.	8.20 C.T.
2381		Cochabamba	,,	18.30			Ollague	,, ,,	14.44	14.20
3854		Viacha	Thursday & Monday		13.08		Calama	,, ,,	21.41	22 00
3700		La Paz	,, ,,	14.25			Antofagasta	Thursday & Sunday	6.23	

Excellent dining and sleeping car service.

PASSENGER FARES FROM ANTOFAGASTA

	To Calama	To Uyuni	To Oruro	To La Paz
1st. clase	£ 0.18. 8	£ 2.11. 4	£ 4. 1. 5	£ 5.12.10
Bed tickets	,, 0.15. 5	,, 0.15. 5	,, 1.10. 2	,, 1.10. 2
Excess luggage (per 100 ks.)	,, 0.14. 0	,, 1.12. 4	,, 2. 3 10	,, 2.19. 4

For local train service and Chilian Northern Railway timetable see other side

Daily Passenger Train to the Nitrate Zone, Calama and Chuquicamata

Altitude above sea level	STATIONS	DAILY		STATIONS	DAILY
		Depart			
	Antofagasta	8.00	CHUQUICAMATA BRANCH	Chuquicamata Depart	6.50 / 16.05
694	Prat	10.40		Calama Arrive	7.43 / 16.58
1027	Baquedano	12.30			Depart
1299	Carmen Alto	13.43		Calama	8.00
1350	Salinas	14.01		Rio Loa	8.16
1360	J. S. Ossa	14.11		Sierra Gorda	10.14
1374	Peineta	14.19		La Noria	10.34
1388	Central	14.29		Solitario	10.52
1405	Maipú	14.41		Placilla	11.07
1417	Unión	14.54		Unión	11.23
1434	Placilla	15.07		Maipú	11 34
1471	Solitario	15.22		Central	11.46
1538	La Noria	15.41		Peineta	11.56
1623	Sierra Gorda	16.10		J. S. Ossa	12.04
2240	Rio Loa	18.25		Salinas	12.15
2272	Calama Arrive	18.41		Carmen Alto	12.37
				Baquedano	13.45
2272	CHUQUICAMATA BRANCH, DAILY — Calama Depart	5.15 / 8.40		Prat	14.59
2677	Chuquicamata Arrive	6.29 / 9.53		Antofagasta Arrive	17.14

Excellent dining car service between Antofagasta & Calama

Direct Trains in Connection with Santiago and Iquique

Santiago	Saturday	Depart 20 30	Iquique	Saturday	Depart 10.45
P. Hundido	Monday	« 17.30	Pintados	»	» 21.30
Catalina	Tuesday	« 1.35	Toco	Sunday	» 4.40
Baquedano	«	« 14 10	Baquedano	»	» 14.00
Toco	«	« 22.50	Catalina	Monday	« 1.00
Pintados	Wednesday	Arrive 5 10	P. Hundido	»	Arrive 7.30
Iquique	«	» 15.00	Santiago	Wednesday	» 11.00

Direct Train between Antofagasta and Pintados, connecting with Nitrate Railway to and from Iquique

Antofagasta	Friday	Depart 9.00	Pintados	Wednesday	Depart 21 30
Baquedano	«	« 14.10	Toco	Thursday	» 4.40
Toco	«	« 22.50	Baquedano	»	» 12.45
Pintados	Saturday	Arrive 5.10	Antofagasta	»	Arrive 16.22

Dining and sleeping car service.

↑ 1924

Market scene at Parotani station on the Cochabamba branch. FCAB

Frontage of Antofagasta station circa 1925. D. Binns collection

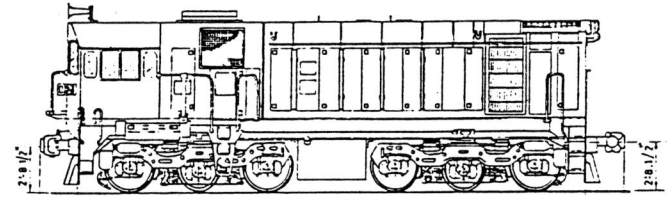

RECENT YEARS – CONTRACTION AND WITHDRAWAL

In 1948 the FCAB commenced operating the newly completed Chilian section of the Northern Transandine Railway on behalf of the Chilian State Railways. This line runs for 112 miles from Augusta Victoria on the Boquete branch, to Socompa on the Argentinian border. Until comparatively recent times traffic on this branch was light, mostly cattle and very few passengers, however, recent returns show a healthy increase in tonnage over the line. The Argentinian section of this line from Salta to the border is interesting, reaching 14,682ft above sea level at Chorrillos Pass. Owned by the Ferrocarriles Argentinos (Argentine State Railways), the line is known simply, to a generation of Argentinian railwaymen, as "Ramal C14". The 250 mile line was started in 1922 and completed in 1948, and to build this spectacular line, the Argentinians employed virtually every trick of the trade in mountain railroading, including two switchbacks and two complete loops. Motive power was drawn from forty-six 2–10–2s supplied by Baldwin and Henschel in 1920–30s and fifteen heavier 2–10–2s built by Skoda in 1949.

On 1 May 1961 the Chilian Government terminated the operating agreement for the Chilian Longitudinal Railway and created a new instrumentality to run the State lines north of the Chanaral Railway, which was named the Ferrocarril Iquique–Pueblo Hundido. This railway incorporated the Chilian Northern Railway, the Ferrocarril de la Provincia de Tarapaca (the metre gauge extension of the Longitudinal line from Pintados to Iquique), and the Ferrocarril Salitrero de Tarapaca, the standard gauge system formerly owned by the London based Nitrate Railways Co. Ltd.

A second contraction of the empire had commenced in February 1959 when the company was obliged to suspend the bulk of its operations in Bolivia and to withdraw expatriate staff. Relationships between the Government and Company had soured following the socialist revolution of 1952. On 3 May 1952 the Rail Brotherhood had issued a manifesto demanding nationalisation of all private railways – without indemnity. Following a union demand that the FCAB Personnel Director be deported as an undesireable alien, radicals stoned the Director's home forcing him to take refuge in the British Embassy! Britain and Bolvia have always had a strained relationship, particularly since an incident in the 1860s when the then rather eccentric President dealt the British Ambassador the ultimate insult. This reached the ears of Her Britannic Majesty, Queen Victoria, who called for an atlas and with a blue pencil struck Bolivia from the map! The President's reaction to this was to mobilise his army and order them to march on England! It is reported that Queen Victoria was not amused . . . !

For the six years following the 1952 revolution operations in the country were difficult. Apart from the threatened expropriation, tariffs were frozen, inflation was rampant, traffic fell dramatically, and attempts by the Company to economise, in light of a collapsed economy, were thwarted by Government decree which compelled the Company to retain in its employment 30% more staff than required to handle the depressed tonnage. Strikes occurred with increasing frequency, there was an outbreak of violence, and the union persisted in attempts to intimidate and interfere with management. When faced with a deficit in 1958 of over half a million pounds sterling, the Company made the decision to withdraw and cut its losses. This it did in February 1959 although it continued to operate the section from Ollague to Oruro.

With the bulk of the last private railway now in local hands the Bolivian Government allowed tariffs to increase by 25% but incompetence and bad management over the next two years all but closed the rail system. The stock of working locomotives within the country was reduced from 65 to 13 and even those were incapable of pulling their tonnage ratings. It was also taking four weeks for freight to move from the border to La Paz! Following a change of Government (this frequency earning Bolivia its unassailable position in the Guiness Book of Records) the incoming President, Victor Paz Estenssoro, in May 1962 sought the assistance of the Company and requested that it return. This offer was rejected, however, the Company signed a two year management agreement covering its line, thus allowing the World Bank time to make a study into the rehabilitation of the entire Bolivian rail system. On 6 October 1964 all FCAB holdings in Bolivia officially passed into the State's possession. Compensation amounting to Stg£2.5 million was not finalised until December 1967.

The last steam locomotives ordered for the railway entered service in 1950 and 1954–55 and were the first to be ordered since 1929. The earlier Garratts had been successful and resulted in a repeat order from Beyer Peacock in 1950 also intended for use in Bolivia. They were an updated version of the earlier design, the most notable visible difference being the fitting of streamlined front and rear tanks. A further order of the standard 2–8–2 design came from Yorkshire Engine Co. in 1955 and were allocated to the Chilian Northern line. By far the most notable of the post-war types were a series of 4–8–2s built by the Vulcan Foundry in 1954–55 and fitted with 19in × 26in cylinders, 4ft coupled wheels and weighing in at 146 tons in working order. They were an extremely handsome design with streamlined style which included smoke deflectors of all things! These latter additions seem curious considering the slow speeds encountered and appear to have been unnecessary and were quickly removed when in service. The engines did however retain their attractive copper capped chimneys and stream-style dome and sand boxes on top of the boiler. A total of sixteen locomotives were provided, ten for Bolivia and six for Chile and with their massive boiler were well suited to the heavy demand for steam on the constant grades of the FCAB lines. These engines were well known for their fine performances on the International passenger trains, but they also handled freight as well.

Steam was still in use in the early 1970s and a photo taken in

Mejillones yard in 1978 shows a long line of serviceable North British 2−8−4Ts and Vulcan Foundry 4−8−2s and it is interesting to note that some of the 2−8−4Ts were still performing shunting duties fitted with Giesl ejectors. These locomotives were cut up in 1982 when the company fell under Chilian control. Steam seems to have been slightly more active in Bolivia during this period, but the appalling conditions under which they had to work led to a similar demise as their Chilian cousins in the 1980s.

Diesel traction was first purchased in 1957 when four Davenport 44 ton locomotives of 1948-build were acquired secondhand from the American Railroad of Puerto Rico. Three of these units are still in service. The last newly built diesel purchased was in 1977 when a single G18U−No. 954−arrived. In 1989 the FCAB purchased from Canadian National a considerable amount of 3ft 6in gauge equipment from the abandoned Newfoundland Railway. Apart from several thousand tons of rail, there were 100 bulkhead flats in the 11600 and 12100 series, and 10 locomotives, GM NF210, (1200hp Co-Co 12 cylinder 567C engine) built between 1956 and 1960. At this time it is not known how many of these locomotives will be regauged for service or cannibalised for spares. Apart from the initial Davenports, all FCAB diesel motive power is General Motors built.

Following the 1964 takeover of the FCAB's Bolivian lines the Company's financial position deteriorated through lack of new investment. In the early 1970s the Chilian Government of President Salvador Allende established a Technical Commission of the Traction and Permanent Way/Works Departments of the State Railways to value the FCAB to form a basis for compensation should the Chilian State decided to nationalise the railway. This State Commission valued the line and assets at Stg£3,600,000; the Company's assessment was £8,000,000! The railway has now established itself as a "bridge route" connecting Antofagasta with Bolivia and Baquedano (the crossing point on the State's Longitudinal Railway) with Argentina utilising the Augusta Victoria arm of the old Boquete branch, the balance of this branch having been abandoned. Reinvestment in the railway commenced in 1981 and a modernisation programme instigated. Track is now being renewed at a rate of 10−12 miles a year and in 1986 a VHF radio train control system came into operation with all locomotives, including yard shunters, so equipped. In 1985 there were 21 diesel locomotives 10 diesel railcar sets, 14 passenger coaches and 2,400 freight wagons listed in use. Freight carried is in the vicinity of 1.3 million (metric) tonnes per annum whilst 10,000 passengers are carried.

On 7 April 1982 Antofagasta Holdings P.L.C. was incorporated as a mirror-image holding company. Under a Scheme of Arrangement all the issued capital of the Antofagasta (Chili) and Bolivia Railway Co. P.L.C., and its subsidiaries, was acquired by means of a one for one share swap. This scheme became effective on 2 July 1982. On 17 November 1982 the Company transferred its residence to Chile with the consent of H.M. Treasury, however, the Company's ordinary and preference shares, and the perpetual debenture stock of the new wholly owned subsidiary, Antofagasta (Chili) & Bolivia Railway Company P.L.C. continued to be listed on the London Stock Exchange.

In the first half of 1982 a controlling interest consisting of 76.09% of voting rights passed from the Chilian owned Turismo a Inmobiliaria Bio-Bio S.A. to the similarly owned Compania de Inversiones Adriatica S.A. On 14 July 1982 the last British chairman, The Viscount Montgomery of Alamein, C.B.E., resigned to be replaced by Mr. A. Luksic and the Board is now resident in Chile. At this time the Company's extensive collection of photographs and memorabilia dating back to the turn of the century was also transferred to Chile to join a growing collection in the Company's newly established private museum in the railway headquarters at Antofagasta.

Today, a century later, the FCAB is a modern dieselised railway system that, despite the hardships of running a railway in an ever changing world, is coping remarkably well. It is however a far cry from the small 2ft 6in gauge Baldwin 2−8−0s which made a name for themselves struggling over steep grades to establish the mining interest, or the sight and sound of four handsome Vulcan 4−8−2s, two at the front and two at the rear, as they heaved the 554 tons of 22 coaches and vans of Train No. 4, the "Internacional" up the 1in33 grade out of La Paz to the Altiplano.

GM diesel at Ollague (altitude 3695 metres) in March 1987. No. 1411 (2nd) is a type GR12U built in 1961 and obtained second-hand from the Chilian State Railways in 1962. The main body colour is a deepish orange with black flashes lined with narrow yellow stripes. The roof is a creamish white and the bogies, steps, underframe detail and end pilots are black.
I. Thomson

5. F.C.A.B. Diesel Locos
LOCOMOTIVE LIST

Running Numbers	Type	H.P.	Model	Builder	Works Numbers	Year Built	In Service	Remarks
600	B–B	380	44 ton	Dav	3046	1948	1958	Ex. American Railroad of Puerto Rico, No. 601.
601–603	B–B	380	44 ton	Dav	3050–3052	1948	1965	Ex. American Railroad of Puerto Rico, No. 605–607. 603 rebuilt 1980 with 500–460hp engine.
950–952	B–B	890	GA8	GM–EMD	28546–28548	1964	1965	
953	B–B	1000	GA18	GM–EMD	34117	1969	1969	
954	A1A–A1A	1000	G18U6	GM–EMD	758010–1	1976	1977	
1400–1405	C–C	1310	GR12U	GM–EMD	26607–26612	1961	1961	
1406–1409	C–C	1310	GR12UD	GM–EMD	28549–28552	1964	1964	
1410	C–C	1500	G22CU	GM–EMD	34118	1969	1969	
1411	C–C	1500	G22CU	GM–EMD	37423	1971	–	Sold to CODELCO (Chilian State Copper Mines) as their number 94 without having worked on FCAB.
1411 (2nd)	C–C	1310	GR12U	GM–EMD	26905	1961	1962	Ex. FCE (Chilian State Railways) Dt 13.014, previously numbered Dt 12.014.
1412	C–C	1310	GR12U	GM–EMD	26908	1961	1962	Ex. FCE (Chilian State Railways) Dt 13.017, prevously numbered Dt 12.017.
	C–C	1200	NF210	GM-Canada		1956–60		Ex. Canadian National 3ft 6in gauge Newfoundland Railways. 38 units originally supplied to CN between 1956 and 1960 and numbered 909–946. No's 910, 911, 915, 916, 921, 926, 936, 944, 945 shipped 1989. No. 933 shipped 1990. (10 in all).

The solitary A1A–A1A G18U6 No. 954 was photographed with caboose No. 237 in tow. I. Thomson

Above: 380hp centre-cab B-B diesel No. 603 is one of a batch of four which were the first diesels on the FCAB. Originally built by Davenport in 1948 for the American Railroad of Puerto Rico, the first (No. 600) was purchased in 1958 followed by No's 601–603 in 1965. L. Russell

Right: Two GM 14XX class diesels climb out of Antofagasta on 22 February 1988 with a typical FCAB freight. The tank wagons are probably bound for Chuquicamata. I. Thomson

FCAB NAME LISTING

R. Stephenson supplied four batches of engines. The Company diagram book, and Livesey, Son & Henderson list give 1883 as the build date for the latter two batches. Stephenson gives 1887/8. Names not known for the first two batches.

Original No's	Wheels	Works No's	Year Built
1–12	4–6–0	2291–2302	1876
18–19	4–2–4–2T	2449–2450	1884

No's 18 and 19 were Webb Compounds, not successful, soon rebuilt with new frames and cylinders, to same specification as original 12.

Original No's	Name	Wheels	Works No's	Year Built
23	Pulacayo	4–6–0	2622	1887
24	Cerrillos	4–6–0	2623	1887
25	Belisario Pero	4–6–0	2624	1887
26	Mariano Ramirez	4–6–0	2633	1888
27	Ascotan	4–6–0	2634	1888
28	Ubina	4–6–0	2635	1888
29	Calama	4–6–0	2636	1888

Of the above 21 engines 12 were still running when the FCAB took control, they were renumbered into the 51–62 block. Builders numbers are not shown as cannibalisation and plate swapping may have occurred. Due to incorrect build dates the diagram book cannot be relied upon.

51	Antofagasta
52	Sucre
53	Cochabamba
54	Ramirez
55	Ascotan
56	Pulacayo
57	Cerrillos
58	Belisaro Pero
59	Vergara
60	Aniceto Arca
61	Ubina
62	Calama

Balance of name listing is an approximate date sequence, original number/1908 number. Only known named engines shown, wheel arrangement as built.

Original No.	1908 No.	Name	Wheels	Builder	Works No.	Year Built
15	1	A. Edwards	0–6–2T	AE	1195	1877
17	2	Don Miguel	2–6–2T	SS	3032	1882

renamed Saldias when renumbered to 2

20	33	Concha y Toro	2–4–2	BLW	8215	1886
21	42	Huanchaca	2–4–2	Rogers	3713	1887
22	31	Rayo	0–4–4T	Rogers	3709	1887

renumbered 33 prior to 1903

30	71	Ollague	2–6–0TT	BLW	9846	1889
31	72	San Pedro	2–6–0TT	BLW	9852	1889
32	82	San Pablo	2–6–0TT	BLW	9855	1889
33	83	Loa	2–6–0TT	BLW	9864	1889
34	84	Rio Grande	2–6–0TT	BLW	9859	1889
35	95	Sierra Gorda	2–8–0TT	BLW	9773	1889
36	3	Relampago	0–6–2ST	BLW	9770	1889
37	81	Bolivar	2–6–0TT	BLW	10469	1889
38	77	San Martin	2–6–0TT	BLW	10470	1889
39	76	Linares	2–6–0TT	BLW	10464	1889
40	–	Olausta	2–4–2	BLW	10942	1890
41	34	Thomas Frias	2–4–2	BLW	10943	1890
42	35	Santa Cruz	2–4–2	BLW	10944	1890
43	73	Tarija	2–6–0TT	BLW	10984	1890
44	74	Tupiza	2–6–0TT	BLW	10988	1890
45	75	Beni	2–6–0TT	BLW	10997	1890
46	–	Hormiga	0–6–2ST	BLW	10998	1890

To Huanchaca Mining 1892 renamed Union

| 47 | 4 | Abeja | 0–6–2ST | BLW | 10995 | 1890 |

Some lists show spelling Aveja

48	78	Lipez	2–6–0TT	BLW	11426	1890
49	79	Chorolque	2–6–0TT	BLW	11436	1890
50	80	San Vincente	4–6–0TT	BLW	11437	1890
51	7	Hormiga	0–6–2ST	BLW	12752	1892

Upon arrival in Chile this engine was renumbered 46 to fill gap vacated when 10998 transferred to F. C. Uyuni–Pulacayo. When shipped from Baldwin named Ardilla

| 52 | 8 | Ardilla | 0–6–2ST | BLW | 12753 | 1892 |

Renumbered 51 upon arrival in Chile. When shipped from Baldwin named Oruga

| 53 | 9 | Vicuña | 0–6–2ST | BLW | 12754 | 1892 |

Renumbered 52 upon arrival in Chile.

53	96	Colquechaca	2–8–0	BLW	12635	1892
54	97	Challapata	2–8–0	BLW	12633	1892
55	98	Sevaruyo	2–8–0	BLW	12667	1892
56	10	Alpaca	0–6–2ST	BLW	14220	1895
57	11	Veloz	0–6–2ST	BLW	14221	1895
58	92	Jupiter	2–8–0	BLW	14461	1895
59	93	Volcano	2–8–0	BLW	14462	1895
60	94	Venus	2–8–0	BLW	14463	1895
61	43	Apolo	4–4–0	BLW	14464	1895
62	44	Minerva	4–4–0	BLW	14465	1895
63	91	Carnot	4–8–0	Cail	2466	1898
64	99	Oruro	2–8–0	BLW	17461	1900
65	100	F. Argandoña	2–8–0	BLW	17462	1900
66	–	Porvenir	2–8–0	Rogers	5544	1900
67	101	Underdown	2–8–0	BLW	18388	1900
68	102	A de Urioste	2–8–0	BLW	18389	1900
69	103	Victoria	2–8–0	BLW	18390	1900
70	104	Trabajo	2–8–0	BLW	18391	1900

Renamed Union, probably when renumbered circa 1907/8

71	105	Polapi	2–8–0	BLW	19437	1901
72	106	Adelante	2–8–0	BLW	19438	1901
73	107	Exito	2–8–0	BLW	19439	1901
74	108	Reserva	2–8–0	BLW	19440	1901
75	109	Boquete	2–8–0	HE	888	1906
76	110	Polpana	2–8–0	HE	889	1906
77	111	Conchi	2–8–0	HE	890	1906
78	112	Uyuni	2–8–0	HE	891	1906
1	12	Chile	0–6–4T	HE	878	1905
2	13	Valdivia	0–6–4T	HE	879	1905
3	14	Mejillones	0–6–4T	HE	880	1905
4	15	Carmen Alto	0–6–4T	HE	907	1906
5	16	Caracoles	0–6–4T	HE	908	1906
7	17	Chuquicamata	0–6–4T	HE	909	1906

8	18	Collahuasi	0–6–4T	HE	910	1906
11	19	Carcote	0–6–4T	HE	911	1906
12	20	Chiguana	0–6–4T	HE	912	1906
79	113	Portezuelo	2–8–0	BLW	27995	1906
80	114	Cuevitas	2–8–0	BLW	28029	1906
81	115	Salinas	2–8–0	BLW	28030	1906
82	116	Central	2–8–0	BLW	28276	1906
83	117	Cere	2–8–0	BLW	28277	1906
84	118	Cebollar	2–8–0	BLW	28282	1906
55	5	Lastenia	0–6–2T	HC	782	1906
56	6	Anita	0–6–2T	HC	783	1906

At this point circa 1907/8 the roster was renumbered. It would appear that tank engines were numbered in the 1–26 blocks, 4 coupled and the 2 Meyers in the 27 to 44 blocks, with 6 and 8 coupled tender engines from 45 upwards

1908 No.	Name	Wheels	Builder	Works No's	Year Built
119	Julaca	2–8–2	HL	2674	1907
120	Chita	2–8–2	HL	2675	1907
121	Quehua	2–8–2	HL	2676	1907
122	Huari	2–8–2	HL	2677	1907
123	Pazña	2–8–2	HL	2678	1907
124	Poopó	2–8–2	HL	2679	1907
125	Machacamarca	2–8–2	HL	2680	1907
126	Potosi	2–8–2	HL	2681	1907
127	La Paz	2–8–2	HL	2682	1907
128	Santiago	2–8–2	HL	2683	1907
129	Ausonia	2–8–2	HE	922	1907
130	Riviera	2–8–2	HE	923	1907
131	Filomena	2–8–2	HE	924	1907
132	Carmen	2–8–2	HE	925	1907
133	Luisis	2–8–2	HE	926	1907
134	Candelaria	2–8–2	HE	927	1907
135	Florencia	2–8–2	HE	928	1907
136	Aconcagua	2–8–2	HE	929	1907
137	Aurelia	2–8–2	HE	930	1907
138	Celia	2–8–2	HE	931	1907
139	Leonor	2–8–2	HC	787	1907
140	Maria	2–8–2	HC	788	1907
141	Reina	2–8–0	HE	958	1908
142	Pelagia	2–8–0	HE	959	1908
143	Carlota	2–8–0	HE	960	1908
144	Alesa	2–8–0	HE	961	1908
145	Mercurio	2–8–0	HE	962	1908
146	Diana	2–8–0	HE	963	1908
147	Flora	2–8–0	HE	964	1908
148	Neptuno	2–8–0	HE	965	1908
149	Saturno	2–8–0	HE	966	1908
150	Magallanes	2–8–0	HE	967	1908
21	Cochrane	0–6–4T	HE	945	1907
22	Cobija	0–6–4T	HE	946	1907
23	Peineta	0–6–4T	HE	947	1907
24	Placilla	0–6–4T	HE	948	1907
25	Latorre	0–6–4T	HE	949	1907
26	Uribe	0–6–4T	HE	950	1907
36	Hércules	2–6–0+0–6–4	K	4534	1908
45	Curico	2–8–0	Alco C	44617	1908
46	San Carlos	2–8–0	Alco C	44618	1908
47	Malleco	2–8–0	Alco C	44619	1908
48	Tarapaca	2–8–0	Alco C	44620	1908
49	Colchagua	2–8–0	Alco C	44621	1908
50	Abra	2–8–0	Alco C	44622	1908
63	El Buitre	2–8–0	Alco C	44623	1908
64	Tacna	2–8–0	Alco C	44624	1908
65	Pampa	2–8–0	Alco C	44625	1908
66	San Salvador	2–8–0	Alco C	44626	1908
67	Concepción	2–8–0	Alco C	44627	1908
68	La Noria	2–8–0	Alco C	44628	1908
69	Atacama	2–8–0	Alco C	44629	1908
70	Canteras	2–8–0	Alco C	44630	1908
85	Coquimbo	2–8–0	Alco C	44631	1908
86	Valparaiso	2–8–0	Alco C	44632	1908
87	Baquedano	2–8–0	Alco C	44633	1908
88	Rio Mulato	2–8–0	Alco C	44634	1908
89	Poderosa	2–8–0	Alco C	44635	1908
90	Santa Rosa	2–8–0	Alco C	44636	1908

F. C. Caleta Coloso a Aguas Blancas
Locomotive names at time of FCAB takeover 24/03/1909

Original No.	Name	Wheels	Builder	Works No.	Year Built
1	B. Dominguez	0–6–2	FCAB	–	1902
2	Laurita	2–8–0	Rogers	5701	1902
3	Zarina	2–8–0	Rogers	5702	1902
4	M. Granja	2–8–0	BLW	24444	1904
5	Tacna	0–6–0T	Hen	6489	1903
6	Carlotita	0–6–0T	Hen	7491	1906
7	Carmen	0–6–0	Hen	6490	1903
8	Iquique	0–6–2ST	Rogers	6270	1905
9	Valparaiso	0–6–2ST	Rogers	6271	1905
10	Barsolona	2–8–0	Alco R	38445	1905
11	Bilbao	2–8–0	Hen	7551	1906
12	Levida	2–8–0	Hen	7753	1906
13	Tarragona	2–8–0	Alco R	41115	1906
14	Gerona	2–8–0	Alco R	41116	1906
15	Galicia	2–8–0	Hen	7754	1906
16	Santiago	0–6–2T	Hen	7550	1906
17	Tocopilla	0–6–0T	Hen	7549	1906
18	H. de Astoreca	0–6–2T	Hen	7958	1907
19	Asturias	0–6–2T	Hen	7959	1907
20	Anita	0–6–0T	Hen	7960	1907
21	Yolanda	0–4–0T	Hen	7995	1907
22	R. Sotomayor	0–4–4–0Tg	Lima	1677	1906
23	Valencia	2–8–0	Hen	8355	1908
24	Andalucia	2–8–0	Hen	8356	1908
25	Aragon	0–6–2T	Hen	8353	1908
26	Baleares	0–6–2T	Hen	8354	1908
27	Nellie	0–4–2			1907

The naming of engines on the FCAB ceased in 1908. All documentation sighted suggests that names were allocated prior to orders been placed on builders. Engines ordered post the 1907/8 renumbering were not allocated names. Names allocated by new owners, after disposal of an engine by the FCAB are omitted, as outside the scope of this narative.

A PICTORIAL RECORD OF THE LAST 25 YEARS OF FCAB STEAM IN BOLIVIA . . .
from the camera of David Ibbotson

With snow capped mountains forming a pleasant background, Alco 2—8—0 No. 401 storms up the hill from La Paz at Km 4 with a short freight train and runs through a shallow cutting topped by the inevitable eucalypt trees.

Bottom left: The crew of Alco 2—8—0 No. 401 pose on the engine as its photograph is taken at the head of a freight train in La Paz yard. The loco has retained its American appearance and would look equally at home on the narrow gauge in Colorado as it does here in Bolivia!

Below: Looking forward along the left hand side of Alco 2—8—0 No. 401 as it stands in the yard at La Paz in February 1955.

Another view of the 10.00am La Paz to Arica train as it climbs up the grade from La Paz to El Alto in February 1955. Also taken from the dining car on the train, this view looks towards the rear of the train and shows Alco 2−8−0 No. 401 providing the very necessary banking assistance despite the fact that the train has powerful Garratt No. 398 at the head.

Bottom right: A contrast in 2−8−0s at Oruro Shed. To the left is American styled No. 404 built by Alco in 1906 whilst to the right is the European styled No. 359 built by Henschel in 1914. Dimensionally the two were similar, but in fittings they were quite different—note the chimneys, domes, smokebox doors and the strengthening bars between smokebox and buffer beam on the Alco.

Below: The front end of Alco 2−8−0 No. 401 at La Paz in April 1954 has a number of interesting features. Notice the Westinghouse brake pump, the bell mounted on the boiler feed, the cover for the chimney, the Rogers name (confirming the loco was built in the Rogers works of Alco) stamped on the cylinder casing and the builders plate, all giving No. 401 a typical American appearance.

Class leader of the Henschel Pacifics built in 1914, No. 333 is seen on a passenger train at Oruro station in January 1956. These graceful engines, which seemed somewhat out of place on the FCAB, were frequent and well liked performers on the passenger trains across the Chilian and Bolivian Altiplano, where the easier grades suited their racy style.

The beauty of the Henschel Pacifics is well illustrated here in this lovely portrait of No. 336 at Oruro in December 1953. Of note is the excellent condition of the loco with its polished brass steam dome placed well back on the boiler and the outside frames, something not normally associated with an express passenger engine!

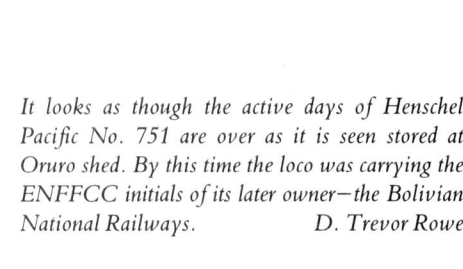

It looks as though the active days of Henschel Pacific No. 751 are over as it is seen stored at Oruro shed. By this time the loco was carrying the ENFFCC initials of its later owner—the Bolivian National Railways.   D. Trevor Rowe

Right: Beyer Garratt No. 390 was the first of the 4–8–2+2–8–4s built in 1928 and the low tanks at each end seem to accentuate the length of the engine. It was seen on a freight at Oruro in March 1956. When they first arrived on the FCAB the three engines of this class were numbered G1–3 but were soon renumbered 390–392 in their correct numerical sequence in the locomotive fleet.

Right: The engine shed at La Paz with two vastly contrasting locomotives in the yard. To the right is the older Alco 2–8–0 No. 401 built in 1906 for the Bolivia Railway Co. as their No. 1, and to the left is the much larger and more modern Beyer Garratt No. 398 built in 1950 and the last of the later series of Garratts.

Below: Beyer Garratt No. 398 is seen here working hard as it rounds a curve at the head of the 10.00am La Paz to Arica train between La Paz and El Alto in February 1955. This shot, which was taken from the dining car on the train, highlights the curvaceuous nature of this steep climb. Of interest also is the starkness of the surrounding hills and flood control masonry baffles in the dry river to prevent the scouring during flash flooding.

A Beyer Garratt provides the power for the twelve cars of the International passenger train as it tackles the steep grades between La Paz and El Alto. An Alco 2–8–0 provides some valuable assistance in the rear and one can just imagine the sound these hard working engines would have made as they rounded the sharp curve!

Overkill! Two of the Alco 2–8–0s, No's 401 and 403, power an extremely short train of only three tank cars and one boxcar at Km 15 on the climb out of La Paz. No. 401 in the lead could probably have handled this train on its own but for some reason it was decided to give assistance from sister engine 403 banking in the rear.

Above: Alco 2–8–0 No. 402 of 1906 is obviously making a major effort to hoist its train up the steep grade out of La Paz in February 1955 as it passes under the electrified FC Guaqui line.

Top right: El Alto station yard in May 1953 with Kitson 2–8–2 No. 409 nicely posed. Built in 1912 this class had 19in×24in cylinders and 44in driving wheels. They were first rebuilt in 1930 but had a further and more extensive rebuilding in 1939 when 21in×24in cylinders incorporating piston valves were fitted.

Right—second picture down: Another of the Hunslet 2–8–0s, in this case No. 145, performs shunting duties in the station platform in Oruro in January 1956. Note the American Army jeep loaded onto the flat car immediately behind the engine.

Right—third picture down: The Beyer Peacock Meyers feature elsewhere in this volume, however, in this case No. 456 is seen in an interesting rear view from the opposite side and clearly shows the rear chimney arrangement which was a feature of these engines. The loco is shown taking water at Viacha station in October 1955.

Bottom right: Rather attractive 2–8–4T No. 412 built as metre gauge in 1912 by Hunslet is seen here at Uyuni shed in March 1955. This engine shows little evidence of having been rebuilt (unlike other engines from this era) and the only addition appears to be the strengthening bars from smokebox to buffer beam.

Top left: Oruro shunter North British 2–8–0 No. 160 is seen in March 1956 looking a little worse for wear and was probably nearing the end of its life. The missing window louvres have been ingeniously replaced with parts of a suitably modified packing case! These little engines had a decidedly unbalanced appearance without their graceful oil headlamps.

Left—second picture down: Henschel 2–8–0 No. 366 at Uyuni shed in March 1955. Built in 1921 and originally numbered 181 this loco had 20½in×24in cylinders, 44in driving wheels and a large boiler which gave it an unbalanced appearance. All of the Henschel metre gauge 2–8–0s seem to have spent their working life entirely in Bolivia.

Left—third picture down: The drooping cab of Henschel 2–8–0 No. 366 looks like it could do with some repairs! Seen here at Uyuni in August 1955 the blower is on as it pauses between shunting duties.

Bottom left: Potosi shed in March 1955 with a much rebuilt 0–6–2T No. 24. Originally built by Hunslet as an 0–6–4T in 1907 it had 15in×18in cylinders and 36in driving wheels. It is unclear what the reason for conversion from 0–6–4T to 0–6–2T was as it appears they performed reasonably well in the original configuration. Despite its humble duties, the loco is still kept in good condition with polished steam dome.

Below: A delightfully Bolivian rural scene at Tolapalca station on the Cochabamba branch with Beyer Garratt No. 394 on a train from Oruro pausing for water in March 1956. The roof handrails of the second boxcar give it away as a metre gauge conversion of one of the original 2ft 6in gauge cars and its size compares notably with the metre gauge built boxcar ahead of it.

Passenger rolling stock at La Paz in July 1954. In the centre is FCAB Saloon No. A7 examples of which were built by Cravens and Ateliers in Belgium. To the left is a smaller inspection saloon which probably had its origins on the 2ft 6in gauge, and to the right is a Dining Car built by Linke Hoffman belonging to the Bolivian section of the Arica–La Paz Railway.

Bolivian Railway Service Car No. B10 is seen here at La Paz Central station in April 1954 and was built by the Middletown Car Works, Pennsylvania USA. To its right is one of the very attractive vestibuled FCAB wooden clerestory roof sleeping cars which were a feature of the FCAB passenger trains.

The railway side of the station at La Paz with a long low level platform and a massive platform canopy as seen here in February 1955. The station clock tower provides a suitable landmark. The eucalypt trees to the right were of Australian origin and were a feature of the railway scene in Chile and Bolivia.

The magnificence of the station building at La Paz is clearly evident in this view taken from the street side in December 1954. Although only served by one or two passenger trains a day, it seemed a fitting terminal for the "highest capital in the world".

A much rebuilt 2−8−4T No. 553 provides shunting power at Oruro in January 1970. Originally built by Kitson in 1912 for the 2ft 6in gauge, it had 17in ×22in cylinders and 37½in driving wheels and was numbered 27. It is shown here as Bolivian National Railways 553 after major rebuilding by the FCAB which featured shortened side tanks, new cylinders incorporating piston valves, and outside steam admission pipes.

D. Trevor Rowe

This photograph depicts a crossing at the Km 66 Passing siding (dead end) on the Cochabamba branch. In March 1956 Beyer Garratt No. 395 is barely visible as it stands in the level passing siding en route to Cochabamba to allow an Oruro bound train on the down grade on the right to pass. Passengers and locals mill around accompanied by a couple of the local dog population! This type of arrangement was quite common in difficult country where facilities for proper crossing of trains, e.g. loops, did not exist. Using this arrangement the main line climbs but the passing siding remains level, the siding can be either straight or curved. In this case the Garratt has drawn past the siding and then reversed its train in to await the downhill train to pass.

The crew of Beyer Peacock Meyer type 0−6−2+0−6−2TT No. 455 built in 1913 take an interest in the photographer as their charge stands in sunlight in 1956. These engines had 18in×20in cylinders and 44in driving wheels but otherwise were individuals as far as positioning of pumps and other plumbing goes! Note that this engine still retains its rear chimney.

Potosi Station in March 1955 with a passenger train in the main platform. Note the very impressive fence which contrasts sharply with the rather undignified condition of the approach road!

Beyer Garratt No. 395 built in 1950 rests from its labours on the Cochabamba branch at Oruro shed in January 1967. Note the sticky ground, a feature of all sheds where oil burning locos were employed and woe betide the enthusiast who walked around without due diligence!

Stubby little Hunslet 2−8−0 No. 151 simmers away quietly at Oruro shed in March 1956 and despite its age (it was built in 1911), appears in good condition. By this time these locos were employed mainly on shunting duties.

A much rebuilt Hunslet 2−8−0 No. 151 built originally in 1911 as a 2ft 6in gauge loco shunts around under the overhead wires of the FC de Guaqui a La Paz in the yard at El Alto in October 1954. This engine has been rebuilt in similar manner to the North British loco on the rear cover and interestingly retains its large oil headlamp.

The largest non-articulated locomotives operated by the FCAB on either gauge were the attractive 146 tons metre gauge 4-8-2s built by the Vulcan Foundry in 1954, the last steam locos built for the Company. They had 19in×24in cylinders and 48in driving wheels, and lasted to the end of steam operations by which time most had lost their smoke deflectors. A highly polished Vulcan Foundry 4-8-2 No. 350 heads up a train consisting mainly of tank cars at Rio Mulato in March 1955. This engine is shown here in original condition with its smoke deflectors and there is no lack of jacking power on the front buffer beam—there are four—should it be needed in case of derailment!

Some of the shed crew pose with Vulcan Foundry 4-8-2 No. 342 at Uyuni shed in March 1955. This photo highlights the size of these locos. With a length of 70ft, width of 10ft 5in and height of 13ft 4¼in, these metre gauge engines were in size as big as a British 9F 2-10-0 (66ft long, 13ft 1in high) and the FCAB engines at 146 tons in working order compared favourably with the 9F at 139 tons. This class's smoke deflectors were soon removed as unnecessary!

In original condition Vulcan Foundry 4-8-2 No. 346, complete with smoke deflectors as originally supplied, stands in the shed yard in La Paz in April 1955. With 19in×24in cylinders and 48in driving wheels, they were ideally suited to the conditions on the FCAB and were well liked by both crews and management.

Henschel 2-8-0 No. 360 built in 1914 just prior to the First World War had 20in×24in cylinders and 44in driving wheels, was photographed at Cochabamba in March 1972. Still carrying its FCAB number, the loco is lettered ENFFCC for the Bolivian National Railways and was subsequently numbered 617.

Henschel Pacific No. 337 makes a spirited departure from Viacha with the 7.50am La Paz–Sucre passenger train on 4 October 1955. Its 19in×24in cylinders and 58½in driving wheels were common to both the 1914 and 1928 built batches of this type. The polished copper dome was characteristic of all the Henschel Pacifics. The semaphores in the background guard the FCAB and FCALP–the Arica–La Paz Railway, which crossed the FCAB here.

A classic shot of Henschel Pacific No. 336 as it stands at Oruro at the head of the passenger train for La Paz. Passengers mill around on the platform and the crew make their final preparations prior to departure. The small boxcar immediately behind the engine is one of the converted 2ft 6in gauge vehicles which were characterised by their roof handrails.

Beyer Garratt 4–8–2+2–8–4T No. 907 of the Bolivian National Railways is seen on an Oruro to Cochabamba train in January 1970. This is one of the "modern" versions of the highly successful FCAB Garratts and was originally FCAB No. 397.   *D. Trevor Rowe*

An evocative scene of the Rio Mulato shed taken at sunset in March 1955 with the tender of Henschel 2–8–0 No. 365 just visible in the deepening gloom.

BOLIVIA RAILWAY CARRIAGES AND WAGONS
from the collection of C. J. Walker

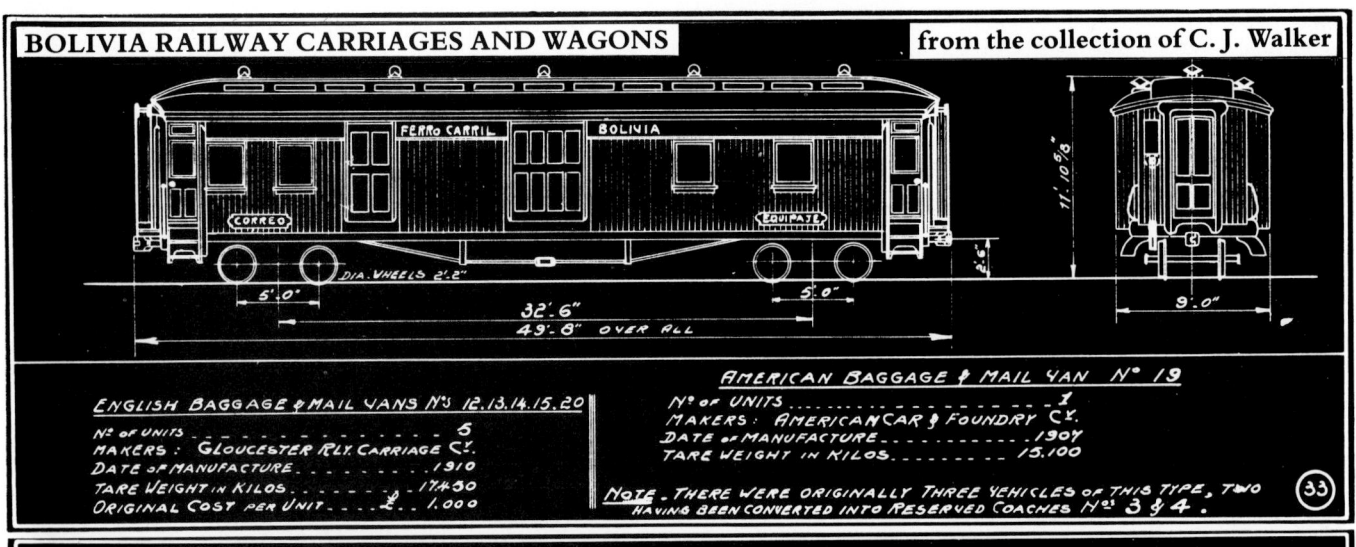

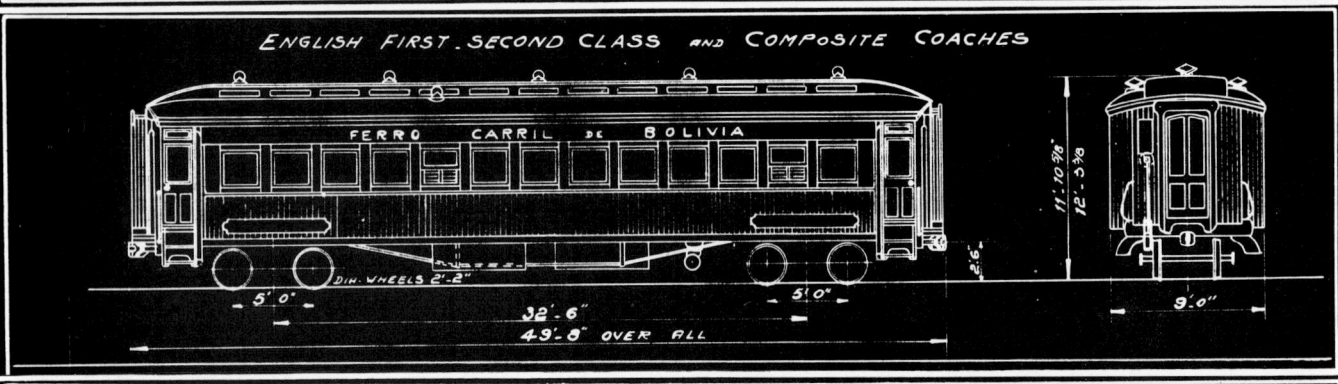

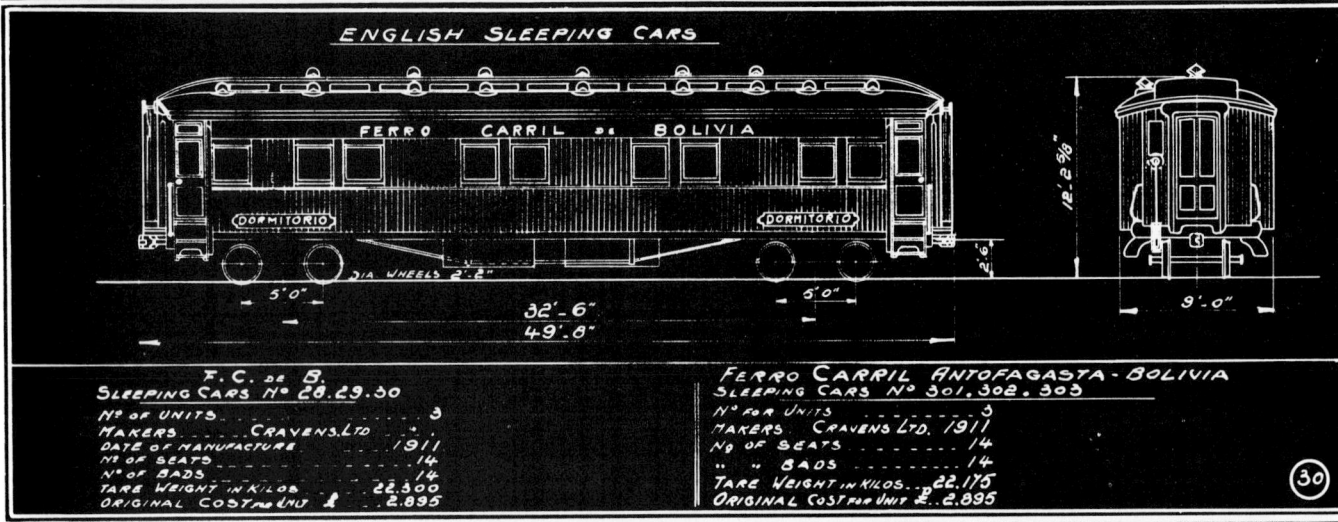

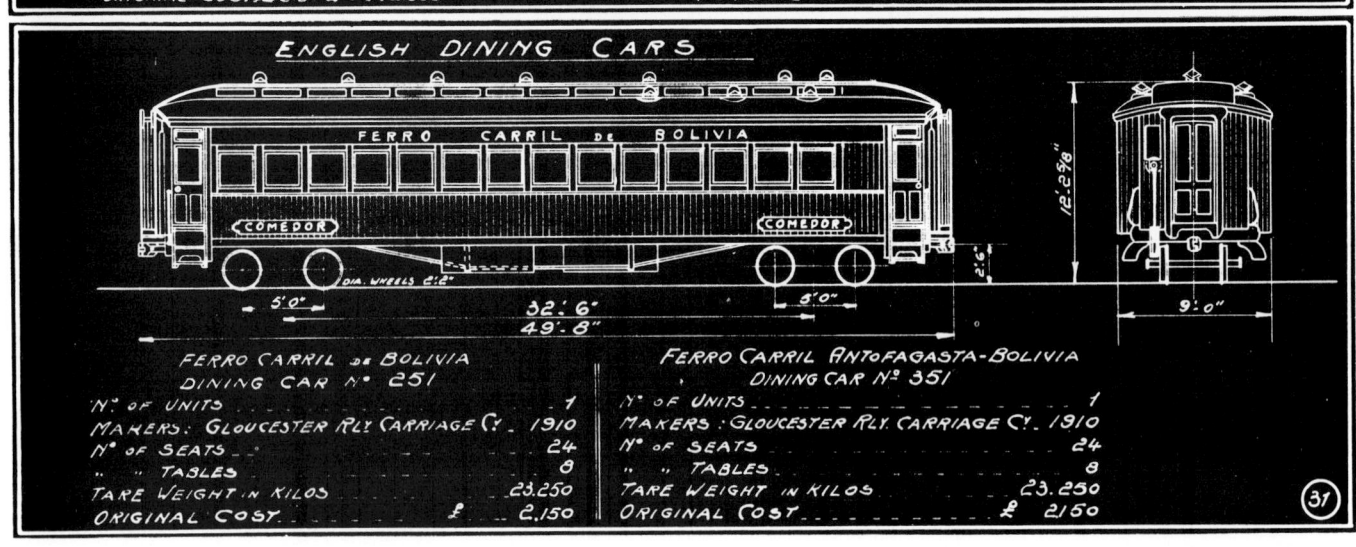

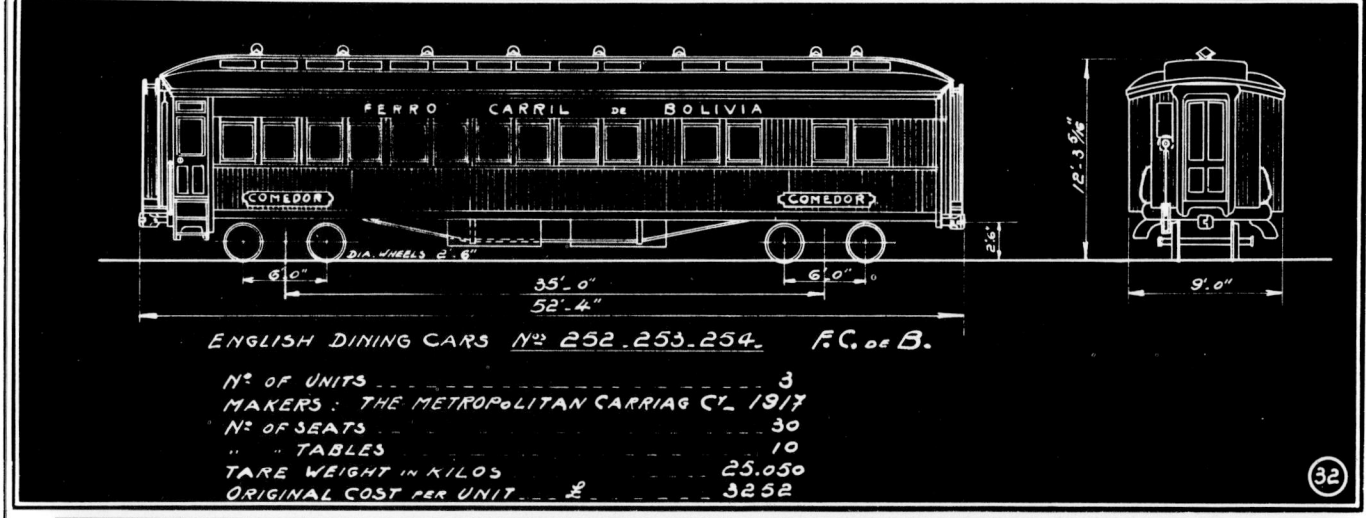

ENGLISH DINING CARS Nº 252.253.254. F.C. de B.

- Nº OF UNITS — 3
- MAKERS: THE METROPOLITAN CARRIAG Cº 1917
- Nº OF SEATS — 30
- " " TABLES — 10
- TARE WEIGHT IN KILOS — 25.050
- ORIGINAL COST PER UNIT — £ 3252

32

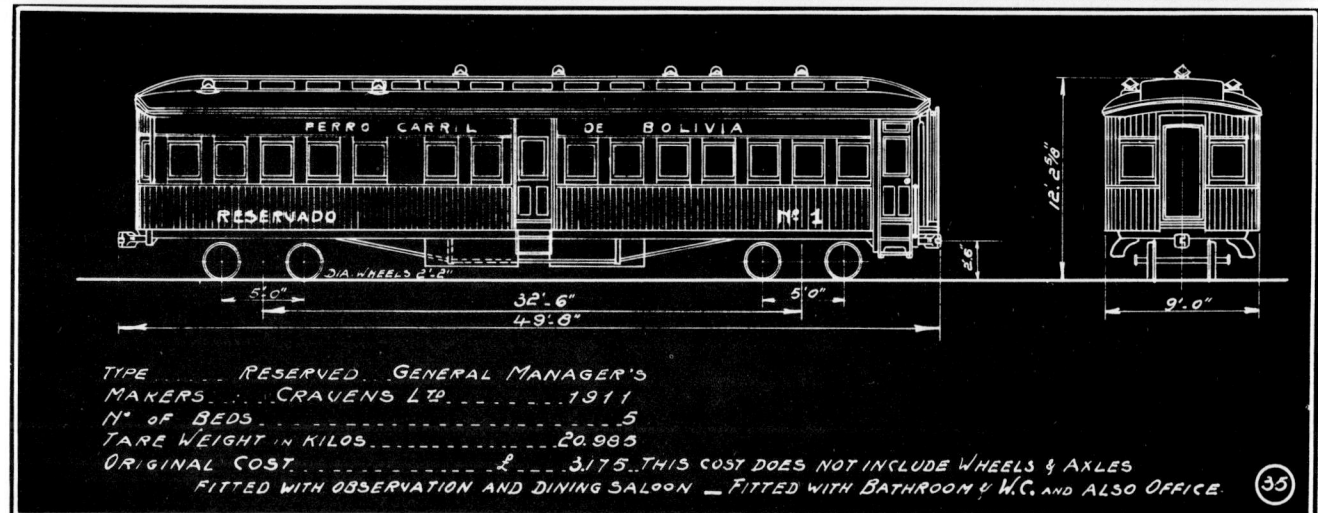

- TYPE — RESERVED GENERAL MANAGER'S
- MAKERS — CRAVENS Lᵗᵈ 1911
- Nº OF BEDS — 5
- TARE WEIGHT IN KILOS — 20.985
- ORIGINAL COST — £ 3.175. THIS COST DOES NOT INCLUDE WHEELS & AXLES
- FITTED WITH OBSERVATION AND DINING SALOON — FITTED WITH BATHROOM & W.C. AND ALSO OFFICE.

35

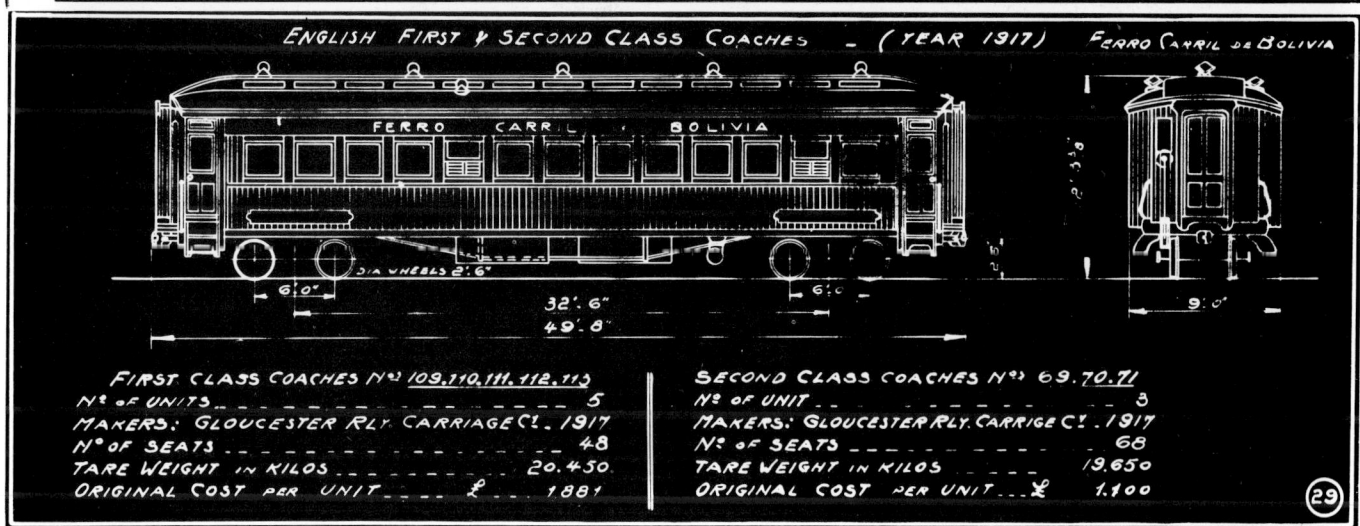

ENGLISH FIRST & SECOND CLASS COACHES — (YEAR 1917) FERRO CARRIL de BOLIVIA

FIRST CLASS COACHES Nºˢ 109,110,111,112,113	SECOND CLASS COACHES Nºˢ 69,70,71
Nº OF UNITS — 5	Nº OF UNITS — 3
MAKERS: GLOUCESTER RLY. CARRIAGE Cº 1917	MAKERS: GLOUCESTER RLY. CARRIGE Cº 1917
Nº OF SEATS — 48	Nº OF SEATS — 68
TARE WEIGHT IN KILOS — 20.450	TARE WEIGHT IN KILOS — 19.650
ORIGINAL COST PER UNIT — £ 1881	ORIGINAL COST PER UNIT — £ 1.100

29

COACHING STOCK   F.C. BOLIVIA RAILWAY

A—	FIRST CLASS	Nº 102	1	UNIT
E—	" "	Nºˢ 103–107	5	"
E—	" "	Nºˢ 108–113	6	"
A—	SECOND "	Nºˢ 54, 55, 57	3	"
E—	" "	Nºˢ 56, 58–62	6	"
E—	" "	Nºˢ 66, 68–71	5	"
A—	COMPOSITE	Nº 203	1	"
E—	"	Nºˢ 204, 205	2	"
E—	"	Nº 206	1	"
E—	SLEEPING	Nºˢ 28–30	3	"
E—	DINING	Nºˢ 251–254	4	"
A—	KITCHEN	Nºˢ 301–302	2	"
A—	RESERVED	Nºˢ 1–11	11	"
A—	CABOOSES	Nºˢ 214, 219, 220	3	"
E—	BAGGAGE-MAIL	Nºˢ 12–15, 20	5	"
A—	" "	Nº 19	1	"

Total   59

A— AMERICAN BUILT
E— ENGLISH BUILT

THE PRICES INDICATED DO NOT INCLUDE COST OF WHEELS AND AXLES IN THE CASES OF THE CAR AND COACHING STOCK.

THE TARES INDICATED FOR THE VARIOUS CAR AND COACHING STOCK MAY NOT BE EXACTLY CORRECT AS SEVERAL IRREGULARITIES HAVE RECENTLY BEEN DISCOVERED. ALL STOCK IS TO BE RE-TARED IF FOUND NECESSARY.

THE DIAGRAMS AND INFORMATION USED IN THIS SECTION WERE EXTRACTED FROM AN F.C de BOLIVIA DIAGRAM BOOK DATED "UYUNI — NOVEMBER 1918". WE WOULD LIKE TO THANK Mr. CHRISTOPHER J. WALKER FOR PROVIDING THIS INFORMATION.

RESERVED CAR Nº. 11. FOR TRAFFIC SUPERINTENDENT. MIDDLETOWN CAR Cº 1906 CONVERTED 1916 FROM CABOOSE Nº 215. TARE WEIGHT IN KILOS 10,060.

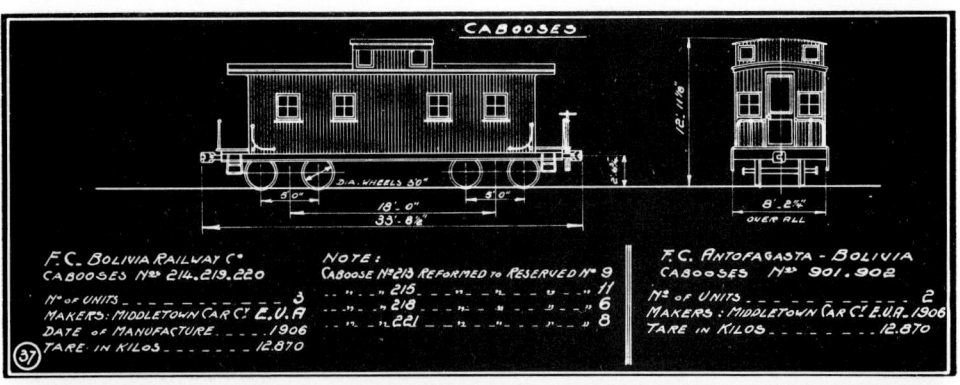

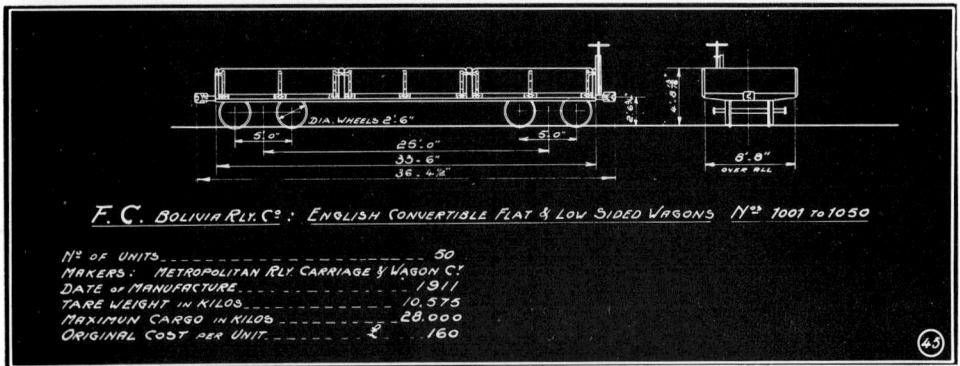

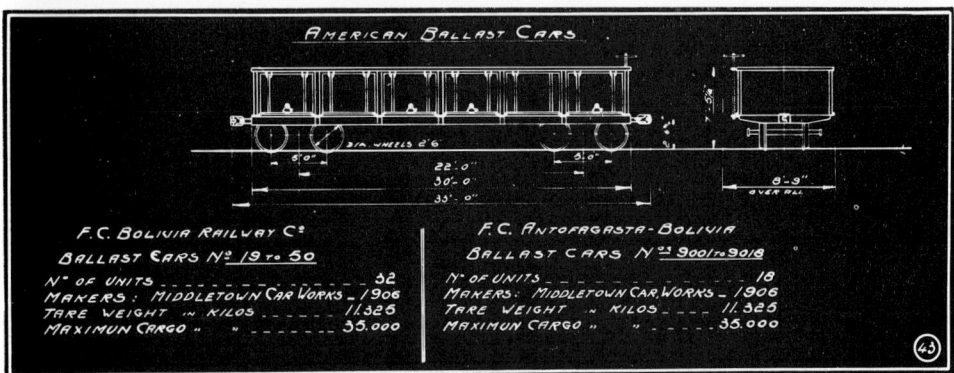

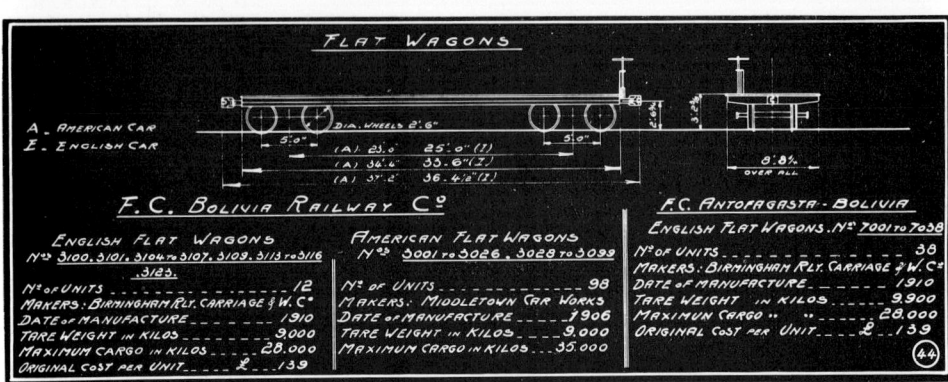

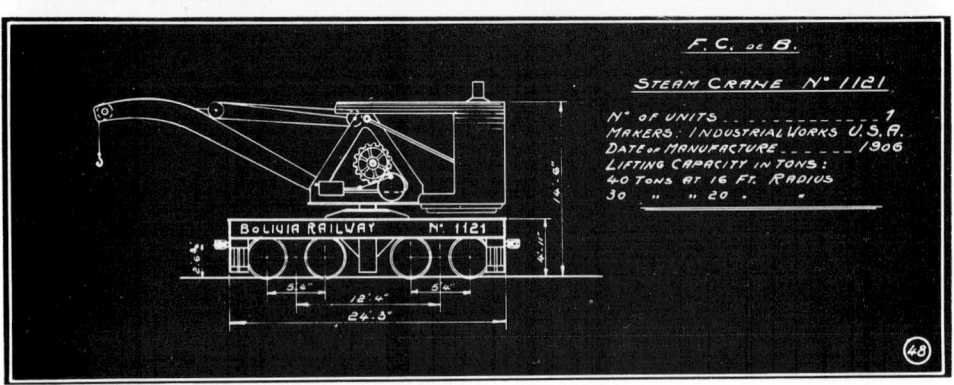

F.C. ANTOFAGASTA – BOLIVIA.

COACHING STOCK NARROW GAUGE IN SERVICE IN BOLIVIA.

BAGGAGE – MAIL	No. 206	1 UNIT.
SECOND CLASS	No. 120	1 "
COMPOSITE	No's 116, 117	2 "
RESERVED	No. 5	1 "

THE FOLLOWING ARE IN SERVICE ON THE INTERNATIONAL TRAIN:

FIRST CLASS	No. 41	1 UNIT.
SECOND "	No's 25, 124	2 "
SLEEPING	No's 16, 17	2 "
COMPOSITE	No. 96	1 "
BAGGAGE – MAIL	No. 222	1 "

F.C. de BOLIVIA.

FIRST CLASS No 102
AMERICAN CAR & FOUNDRY Co. 1907
No. of SEATS   44
TARE WEIGHT IN KILOS   18,000.

SECOND CLASS No's 54, 55, 57.
AMERICAN CAR & FOUNDRY Co. 1907
No. of SEATS   64
TARE WEIGHT IN KILOS   18,000.

FIRST & SECOND COMPOSITE No 203
AMERICAN CAR & FOUNDRY Co. 1907
No. of FIRST CLASS SEATS   20
" SECOND   "   40
TARE WEIGHT IN KILOS   18,000.

RESERVED COACHES No's 2–5, 7, 10.

No. 2. RESERVED – ENGINEERS INSPECTION COACH. AMERICAN CAR & FOUNDRY Co. 1907. CONVERTED FROM COMPOSITE COACH No. 202.
TARE WEIGHT IN KILOS 18,140.

No. 3. RESERVED FOR CHIEF OF DEPARTMENTS. AMERICAN CAR & FOUNDRY Co. 1907. CONVERTED TO RESERVED IN 1914 FROM BAGGAGE – MAIL No. 8.
TARE WEIGHT IN KILOS 18,400.

No. 4. FOR PAYMASTER. AMERICAN CAR & FOUNDRY Co. 1907. CONVERTED TO RESERVED 1914 FROM BAGGAGE – MAIL No. 9.
TARE WEIGHT IN KILOS 16,725.
BY 1918 FITTED UP FOR THE CHANGE OF GAUGE BETWEEN ORURO AND UYUNI.

No. 5. RESERVED FOR SECTIONAL ENGINEER. MIDDLETOWN CAR Co. 1906.
TARE WEIGHT IN KILOS 16,650.

No. 7. RESERVED – PROVISIONS VAN. MIDDLETOWN CAR Co. (USA) 1906. CONVERTED 1916 FROM CATTLE WAGON No. 4007.
TARE WEIGHT IN KILOS 13,500.

No. 10. RESERVED – ACCOUNTS DEPARTMENT. MIDDLETOWN CAR Co. 1906. CONVERTED 1916 FROM CATTLE WAGON No. 4019.
TARE WEIGHT IN KILOS 14,500.

RESERVED COACHES No's 6, 8, 9.

No. 6. RESERVED FOR LOCOMOTIVE SUPERINTENDENT. MIDDLETOWN CAR Co. 1906. CONVERTED 1917 FROM CABOOSE No. 218.

No. 8. RESERVED FOR STORES DEPT. MIDDLETOWN CAR Co. 1906. CONVERTED 1915 FROM CABOOSE No. 221.

No. 9. RESERVED FOR ACCOUNTS DEPT. MIDDLETOWN CAR Co. 1906. CONVERTED 1915 FROM CABOOSE No. 213.
TARE WEIGHT IN KILOS. 14,000.

WAGON STOCK — F.C. BOLIVIA RAILWAY

A — COVERED GOODS No's 2001, 2003-2013, 2015-2028.	26 UNITS	
E — COVERED GOODS No's 2029-2039, 2041-2096.	67 UNITS	
A — CATTLE WAGONS No's 4003-4006, 4008, 4009, 4011-4015, 4017, 4018.	13 UNITS	
E — OPEN GOODS No's 51-60.	10 UNITS	
A — BALLAST CARS No's 19-50.	32 UNITS	
A — FLAT CARS No's 3001-3026, 3028-3099.	98 UNITS	
E — FLAT CARS No's 3100, 3101, 3104-3107, 3109, 3113-3116, 3123.	12 UNITS	
E — CONVERTIBLES No's 1001-1050.	50 UNITS	
E — GUNPOWDER VANS No's 5001-5004.	4 UNITS	
A — WATER TANKS No's 1-8.	8 UNITS	
E — WATER TANKS No's 13-14.	2 UNITS	
A — STEAM CRANE No. 1121.	1 UNIT	
A — SNOWPLOUGH	1 UNIT	
A — REFRIGERATOR CAR No. 401 — CONVERTED FROM COVERED GOODS No. 2002.	1 UNIT	
A — BREAKDOWN CAR No. 4016 — CONVERTED FROM CATTLE WAGON No. 4016.	1 UNIT	
	TOTAL	326

A — AMERICAN BUILT
E — ENGLISH BUILT

FROM A DIAGRAM BOOK "UYUNI NOVEMBER 1918".

F.C. ANTOFAGASTA — BOLIVIA

COACHING STOCK

E — FIRST CLASS	No's 401-404.	4 UNITS
E — SECOND "	No's 451-457.	7 UNITS
F — COMPOSITE	No 501.	1 UNIT
E — SLEEPING	No's 301-303.	3 UNITS
E — DINING	No 351	1 UNIT
A — CABOOSES	No's 901-902	2 UNITS
E — BAGGAGE MAIL	No's 551-553	3 UNITS
		TOTAL	21

WAGON STOCK

E — COVERED GOODS	No's 6001-6082	82 UNITS
A — CATTLE WAGONS	No's 701-702	2 UNITS
E — OPEN GOODS	No's 8001-8040	40 UNITS
A — BALLAST CARS	No's 9001-9018	18 UNITS
E — FLAT CARS	No's 7001-7038	38 UNITS
E — GUNPOWDER VANS	No's 601-606	6 UNITS
E — WATER TANKS	No's 801-804	4 UNITS
		TOTAL	190

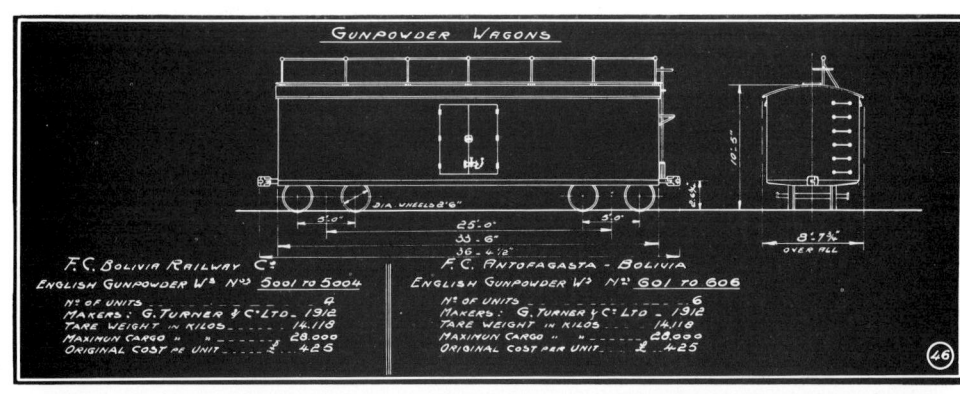

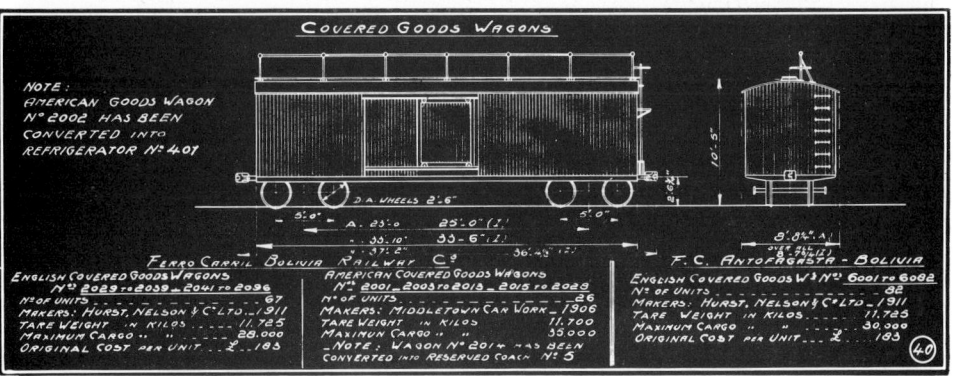

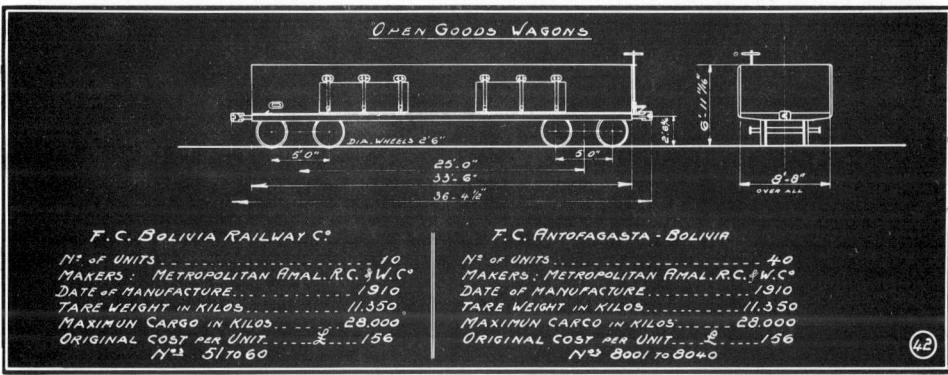

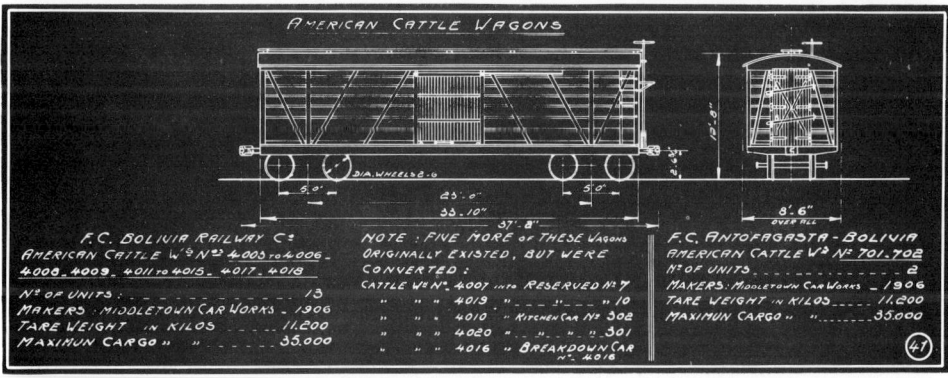

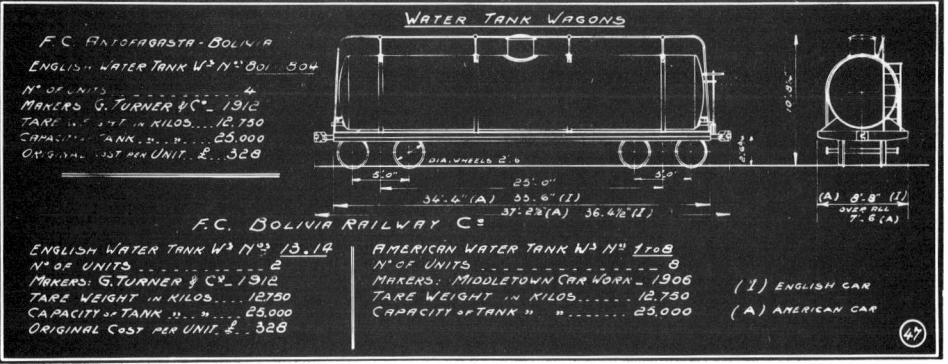

FCAB Distances and other details at February 1912

Distance in km from Antofagasta	Station	Height above sea-level —metres	Length of second track —metres	Length of other tracks —metres	Length of branch line —metres	Capacity of Water Tanks —tons	Capacity of coal stores —cubic metres	No. of Loco-motives	Turn-table
	Antofagasta	—	149	22000	—	130	6290	40	T
3.628	Desvio Lastre	—	—	150	—	—	—	—	—
4.053	Desvio Condensadores	—	—	—	3084	—	—	—	—
4.265	Ramal á Estación Norte	—	—	—	3680	—	—	—	—
4.110	Playa Blanca	60	237	523	—	—	—	—	—
10.760	Desv. Canteras	—	—	167	—	—	—	—	—
9.320	Carrizo	—	130	—	—	28	—	—	—
13.790	Sarj. Aldea	295	357	430	—	28	—	—	—
20.300	La Negra	408	446	422	—	—	—	—	—
29.381	Portezuelo	554	520	2028	—	200	—	—	Y
29.610	R. á Salar del Carmen	—	—	—	6700	—	—	—	—
35.677	O'Higgins	531	450	355	—	—	—	—	—
35.850	Ram. Boquete	—	—	—	111415	—	—	—	—
47.187	Uribe	573	467	—	—	—	—	—	—
58.810	Ram. Mejillones	—	—	—	76630	—	—	—	—
59.065	Prat	682	520	3319	—	245	—	12	T.C.
70.630	Latorre	783	316	—	—	130	—	—	—
83.609	Cuevitas	893	480	565	—	100	—	—	Y
83.560	Des. Barnett	—	—	140	—	—	—	—	—
96.752	Baquedano	1014	452	—	—	130	—	—	—
103.940	Desvio Zuleta	—	—	243	—	—	—	—	—
109.469	Cerrillos	1161	570	527	—	130	—	—	Y
117.345	El Buitre	1231	350	130	—	—	—	—	—
117.373	Of. Riviera	—	—	—	2033	—	—	—	—
120.671	Of. Florencia	—	—	—	4985	—	—	—	—
120.671	Of. Celia	—	—	—	6906	—	—	—	—
120.771	Santa Rosa	1279	148	—	—	—	—	—	—
122.395	Of. Feo. Puelma	—	—	—	557	—	—	—	—
122.478	Carmen Alto	1285	308	338	—	—	—	—	—
128.598	Salinas	1341	334	505	—	420	—	—	Y
128.660	Of. Lastenia	—	—	—	388	—	—	—	—
128.660	Of. Aurelia	—	—	—	3087	—	—	—	—
128.660	Of. Carmela	—	—	—	5633	—	—	—	—
131.694	José S Ossa	1360	320	—	—	—	—	—	—
131.786	Of. J. S. Ossa	—	—	—	741	—	—	—	—
133.892	Peineta	1369	454	200	—	—	—	—	—
133.871	Of. Ausonia	—	—	—	3401	—	—	—	—
133.871	Of. Leonor	—	—	—	7329	—	—	—	—
133.871	Of. Cecilia	—	—	—	7450	—	—	—	—
136.371	Of. A. Edwards	—	—	—	—	—	—	—	—
136.833	Central	1383	597	737	—	—	—	—	Y
140.831	Maipú	1403	730	490	—	—	—	—	—
141.056	Of. A. Pinto	—	—	—	2100	—	—	—	—
141.018	Of. A. Prat	—	—	—	1398	—	—	—	—
144.220	Union	1414	519	300	—	200	—	11	Y
144.287	Of. Anita	—	—	—	1789	—	—	—	—
144.287	Of. Candelaria	—	—	—	3762	—	—	—	—
144.287	Of. Luisis	—	—	—	5734	—	—	—	—
144.287	Of. Angamos	—	—	—	11148	—	—	—	—
148.421	Placilla	1431	295	—	—	—	—	—	—
148.593	Of. Maria	—	—	—	1635	—	—	—	—
148.593	Of. Curicó	—	—	—	5912	—	—	—	—
154.346	Solitario	1470	307	260	—	—	—	—	—
154.272	Of. Filomena	—	—	—	995	—	—	—	—
162.208	Of. Aconcagua	—	—	—	636	—	—	—	—
162.187	La Noria	1534	183	—	—	—	—	—	—

Distance in km from Antofagasta	Station	Height above sea-level — metres	Length of second track — metres	Length of other tracks — metres	Length of branch line — metres	Capacity of Water Tanks — tons	Capacity of coal stores — cubic metres	No. of Loco-motives	Turn-table
171.280	Sierra Gorda	1623	234	1108	—	200	346	1	Y
171.578	Do. Co. Progreso	—	—	—	—	—	—	—	—
179.617	Cochrane	1727	288	771	—	28	—	—	Y
205.878	Cerritos Bayos	2142	164	488	—	45	—	—	Y
239.080	Calama	2265	363	1909	—	200	1932	19	Y
240.698	Do. Chorrillos	—	—	—	1386	—	—	—	—
240.698	Do. Huamachuco	—	—	—	253	—	—	—	—
253.738	Ramal Chuquicamata	—	—	—	10085	—	—	—	—
253.804	San Salvador	2467	76	—	—	117	—	—	—
270.358	Cere	2641	298	—	—	28	—	—	—
300.071	Ramal Abra	—	—	—	19020	—	—	—	—
300.112	Conchi	3015	—	290	—	19	—	—	—

In 1918 a 4.8km deviation was built to avoid the Conchi viaduct which crossed the river Loa. The distances shown are from official figures dated February 1912. For km from Antofagasta 1918 on, add 4.8km from San Pedro forward, i.e. Ollague would become 441.4km.

Distance in km from Antofagasta	Station	Height above sea-level — metres	Length of second track — metres	Length of other tracks — metres	Length of branch line — metres	Capacity of Water Tanks — tons	Capacity of coal stores — cubic metres	No. of Loco-motives	Turn-table
313.395	San Pedro	3223	277	140	—	56	—	—	—
313.400	Do. Represa	—	—	—	401	—	—	—	—
340.905	Polapi	3772	208	—	—	44	—	—	—
361.566	Ascotan	3955	310	1029	—	—	—	—	Y
388.438	Ramal Borax	—	—	—	—	—	—	—	—
388.515	Cebollar	3729	265	446	—	56	—	—	Y
403.820	Carcote	—	330	—	—	—	—	—	—
412.430	San Martin	3692	378	918	—	—	—	—	Y
436.570	Ollague	3696	255	1210	—	56	819	5	Y
436.919	R. Collahuasi	—	—	—	94920	—	—	—	—
471.256	Chiguana	3678	195	—	—	—	—	—	—
517.168	Julaca	3668	239	242	—	57	—	—	—
546.952	Rio Grande	3658	192	—	—	48	—	—	—
611.630	Uyuni	3660	344	5286	—	200	2000	19	T
627.050	Do. La Sal	—	—	—	47	—	—	—	—
687.688	Chita	3745	137	—	—	28	—	—	—
716.355	Rio Mulato	3806	—	316	—	200	—	2	Y
717.166	Do. Cantera	—	—	—	458	—	—	—	—
754.980	Tambo Viejo	—	—	150	—	—	—	—	—
761.835	Sevaruyo	3728	224	166	—	28	—	—	—
802.147	Huari	3699	187	—	—	28	—	—	—
814.255	Challapata	3708	229	945	—	29	—	—	—
815.971	Do. Bosman	—	—	—	250	—	—	—	—
851.547	Pazña	3700	136	73	—	28	—	—	—
871.440	Do. Trinacria	—	—	—	535	—	—	—	—
876.224	R. Bella Vista	—	—	—	3428	—	—	—	—
876.224	R. Alantaña	—	—	—	1424	—	—	—	—
877.380	Poopó	3708	215	186	—	18	—	—	Y
901.170	R. Ing. Machacamarca	—	—	—	2433	—	—	—	—
901.226	Machacamarca	3703	187	259	—	14	—	—	—
925.273	R. á San José	—	—	—	3911	—	—	—	—
925.273	Oruro (F.C.A.B.)	3696	123	1918	—	23	—	3	T

RAMAL MEJILLONES

Distance in km from Antofagasta	Station	Height above sea-level — metres	Length of second track — metres	Length of other tracks — metres	Length of branch line — metres	Capacity of Water Tanks — tons	Capacity of coal stores — cubic metres	No. of Loco-motives	Turn-table
74.810	La Cumbre	966	260	1500	—	100	—	—	Y
87.410	Desesperado	684	275	—	—	100	—	—	—
98.510	Canteras	400	240	192	—	—	—	—	—
107.427	Pampa	217	250	1450	—	100	—	—	Y
110.430	Nivel	163	234	195	—	—	—	—	—
134.850	Mejillones	—	150	12000	—	200	D	14	Y.T

Distance in km from Antofagasta	Station	Height above sea-level —metres	Length of second track —metres	Length of other tracks —metres	Length of branch line —metres	Capacity of Water Tanks —tons	Capacity of coal stores —cubic metres	No. of Loco-motives	Turn-table
			RAMAL ABRA						
319.031	Abra	3490	100	180	—	—	—	—	—
			RAMAL COLLAHUASI						
442.903	D. p. Lastre	—	—	263	—	—	—	—	—
468.455	Cosca	—	210	—	—	—	—	—	—
473.305	Puquios	4163	150	150	—	100	—	—	—
486.905	Yuma	4395	150	—	—	—	—	—	—
507.229	Ujina	4253	300	1427	—	100	D.	—	Y
525.880	Montt.	4703	140	200	—	—	—	—	Y
525.955	R. á M. Grande	—	—	5820	—	—	—	—	—
526.005	R. á M. Poderosa	—	—	640	—	—	—	—	—
526.645	M. Poderosa	—	—	—	—	—	—	—	—
531.825	Punto Alto	4821	—	—	—	—	—	—	—
			RAMAL CHUQUICAMATA						
261.911	Chuquicamata	2694	128	243	—	—	—	—	Y

Distance in km from Antofagasta	Station	Height above sea-level —metres	Length of second track —metres	Length of other tracks —metres	Length of branch line —metres	Capacity of Water Tanks —tons	Capacity of coal stores —cubic metres	No. of Loco-motives	Turn-table
			RAMAL BOQUETE						
68.850	Lata	611	300	30	—	—	—	—	—
88.850	Llanos	726	400	360	—	100	—	—	—
113.850	Cerro Negro	1051	275	—	—	56	—	—	—
138.250	Boquete	1520	275	—	—	260	D	4	Y
142.560	Of. Savona	—	—	—	2879	—	—	—	—
143.450	Of. Pissis	—	—	—	276	—	—	—	—
147.265	Of. Domeyko	1721	—	88	—	—	—	—	Y

Distance in km from Coloso	Station	Height above sea-level —metres	Length of second track —metres	Length of other tracks —metres	Length of branch line —metres	Capacity of Water Tanks —tons	Capacity of coal stores —cubic metres	No. of Loco-motives	Turn-table
			RAMAL AGUAS BLANCAS						
	Coloso	—	250	4000	—	100	11000	15	T
2.—	Y	—	—	—	—	—	1200	—	Y
10.700	Carrizo	—	130	—	—	—	—	—	—
22.315	La Negra	408	150	890	—	180	—	—	Y
52.178	Varillas	862	280	180	—	30	—	—	—
66.300	Laguna Seca	1041	240	—	—	—	—	—	Y
92.440	Of. Pampa Rica	—	—	900	—	—	—	—	—
91.863	Yungay	968	240	1400	—	90	180	—	T
97.883	Of. Pepa	961	—	1200	—	—	—	—	—
106.745	Of. Oriente	969	—	1100	—	—	—	—	—
108.389	Of. Americana	925	—	430	—	—	—	—	—
125.073	Of. Castilla	788	—	1200	—	—	—	—	—
113.161	Of. Avanzada	1042	—	600	—	—	—	—	—
101.263	Of. Eugenia	1020	170	1050	—	—	—	—	—
103.880	Of. Petronila	1020	—	800	—	—	—	—	—
112.897	Of. San Gregorio	1130	—	800	—	—	—	—	—
116.356	Of. Valparaiso	1203	—	1000	—	—	—	—	—
109.040	Of. Maria Teresa	1135	—	800	—	—	—	—	—
111.381	Of. Bonasort	1146	—	900	—	—	—	—	—